THE NAKED FUTURE

CURRENT

THE NAKED FUTURE

WHAT HAPPENS IN A WORLD THAT ANTICIPATES YOUR EVERY MOVE?

PATRICK TUCKER

Current

CURRENT
Published by the Penguin Group
Penguin Group (USA) LLC
375 Hudson Street
New York, New York 10014

USA | Canada | UK | Ireland | Australia | New Zealand | India | South Africa | China
penguin.com
A Penguin Random House Company

First published by Current, a member of Penguin Group (USA) LLC, 2014

LIBRARY OF CONGRESS CATALOGING-IN-PUBLICATION DATA

Tucker, Patrick, 1976–
The naked future : what happens in a world that anticipates your every move / Patrick Tucker.
pages cm
Includes bibliographical references and index.
ISBN 978-1-59184-586-7
1. Information technology—Social aspects. 2. Forecasting. 3. Big data—Social aspects. 4. Technological innovations—Social aspects. 5. Privacy, Right of. I. Title

HM851.T83 2014
005.7—dc23
2013040374

Printed in the United States of America
10 9 8 7 6 5 4 3 2 1

Set in Sabon
Designed by Spring Hoteling

To Beth

CONTENTS

THE NAKED FUTURE

INTRODUCTION

IMAGINE waking up tomorrow to discover your new top-of-the-line smartphone, the device you use to coordinate all your calls and appointments, has sent you a text. It reads:

Today is Monday and you are probably going to work. So have a great day at work today!—Sincerely, Phone.

Would you be alarmed? Perhaps at first. But there would be no mystery where the data came from. It's mostly information that you *know* you've given to your phone.

Now consider how you would feel if you woke up tomorrow and your new phone predicted a much more seemingly random occurrence:

Good morning! Today, as you leave work, you will run into your old girlfriend Vanessa (you dated her eleven years ago), and she is going to tell you that she is getting married. Do try to act surprised!

What conclusion could you draw from this but that someone has been stalking your Facebook profile and knows you have an old girlfriend named Vanessa? And that this someone has probably been stalking *her* profile as well and spotted her engagement announcement. Now this ghoul has hacked your calendars and your phone!

Unsure what to do, let's say you ignore it for the time being. But then, as you're leaving work, the prophecy holds true and you pass Vanessa on the sidewalk. Remembering the text from that morning, you congratulate her on the engagement. Her mouth drops and her eyes widen with alarm.

"How did you know I was engaged?" she asks.

You're about to say, "My phone sent me a text," but you stop yourself just in time.

"Didn't you post something to your Facebook profile?" you ask.

"Not *yet,*" she answers and walks hurriedly away.

You should have paid attention to your phone and just acted surprised.

This scenario is closer to reality than you might think. In fact, the technology and data already exist to make it happen. We give it away to retailers, phone companies, the government, social networks, and especially our own phones without realizing it. In the next few years that data will become more useful to more people. This is what I call the naked future.

The capital-*F* Future was born of the Enlightenment-era notion of progress, the idea that the present—in the form of institutions, products, fashions, tastes, and modes of life—can and must be continually reformed and improved. This is why our interaction with the future as groups and as nations is an expression of both personal and national identity. As a public idea, the future shapes buying, voting, and social behavior. The future is an improved present, safer, more convenient, better managed through the wonders of technology and invention.

But the future—in the form of intention—is also an incredibly private idea. Your future, whether it's what you're going to do tonight, next year, or the next time you've got a thousand bucks to burn, is invisible to everyone but you. We are jealous guards of the personal, secret future, and with good reason. Imagine if any act you were going to commit was laid bare before the world, how naked you would feel.

In the next two decades, we will be able to predict huge areas of the future with far greater accuracy than ever before in human history, including events long thought to be beyond the realm of human inference. The rate by which we can extrapolate meaningful patterns from the data of the present is quickening as rapidly as is the spread of the Internet because the two are inexorably linked. The Internet is turning prediction into an equation. Mathematicians, statisticians, computer scientists, marketers, and hackers are using a global network of sensors, software programs, information collection devices, and apps to reveal in ever-greater detail the effects of our perpetual reform on the world around us. From programs that chart potential flu outbreaks to expensive (yet imperfect) "quant" algorithms that anticipate bursts of stock-market volatility, computer-aided prediction is everywhere.

Big Data Is Dead. Long Live Big Data

Between November 2010 and February 2013, the number of queries related to the term "big data" jumped by a factor of twenty-nine. That means that if big data were a country that grew every time someone searched for it on Google, it would be the size of the United Kingdom in 2010 and the size of Australia just three years later. It's a hot topic, but it's also a phrase that means something different depending on who is trying to sell you what. A couple of years ago, the term referred to data sets so large that the owners of those sets couldn't derive any insight from them. Big data was a euphemism for unstructured and unworkable bits of information locked away in servers, or worse, on paper. This quality of bigness made those little values on spreadsheets effectively valueless. No more. Go to any IT conference today and you'll find rooms full of vendors so eager to work with your big data they will be unable to refrain from shoving flash drives into your pockets. Large companies and the government now work with big data all the time.

On February 16, 2012, the phrase "big data" made an evolutionary leap with the publication of a piece by Charles Duhigg in the

New York Times. The article exposed how the retail chain Target used records of millions of transactions (and information from its baby registry) to draw a corollary between the purchase of various common items such as unscented baby lotion and pregnancy. When Target began sending coupons for baby supplies to customers who it had statistically deduced were in a family way, one customer's father had a fit, demanded an explanation, and realized that a soulless company with a lot of records had discovered something extremely intimate about his daughter before she had had a chance to break the news to him. The story was picked up on *The Colbert Report* and *The Daily Show*, and was repeated on blogs and news stories around the world. Big data went from a boring business idea to a menacing force for evil. It was a secret statistical prescient power that enormous institutions used against the rest of us. The *Guardian* newspaper's 2013 revelations about the scope and power of the NSA to surveil communications among U.S. citizens only added to this narrative. We feel we have arrived at an age in which our devices communicate about us in a language we cannot hear to parties we cannot see. Big data belongs to *them*, not us. We are its victims.

This view of big data is not entirely incorrect. As you'll find in this book, companies, emboldened by new capabilities, are eager to use the enormous data sets they've amassed to squeeze more money out of their present and future customers. Governments, too, are using big data to do more with less, which is fine—as long as you approve of everything the government does.

But the view of big data as a dark force available only to large institutions is limited. Big data will shrink, becoming small enough to fit inside single-push notification on a single user's phone. Most of what we understand about it represents its past, when it was *solely* a capability that the powerful used to gain leverage over the weak. The future of this resource is incredibly open to consumers, activists, and regular people. But big data is only one piece of a larger trend that's reshaping life on this planet and exposing the future.

With very little fanfare, we have left the big data era and have entered the *telemetric* age, derived from the word "telemetry": "The process or practice of obtaining measurements in one place and relaying them for recording or display to a point at a distance. The transmission of measurements by the apparatus making them."[1] Telemetry is the collection and transfer of data in *real time*, as though sensed. If you've ever been in a hospital and had an EKG, ECG, or any sort of monitoring device attached to you, if you've ever been able to see your cardiac activity displayed heartbeat for heartbeat with the knowledge that that data stream was also reaching the nurse down the hall, possibly even your doctor on his smartphone, then you've experienced telemetry. The reach and power of telemetry is what separates the less predictable world in which we evolved our humanity from the more predictable one in which that humanity will grow and be tested.

Telemetry is what divides the present from the naked future.

As sensors, cameras, and microphones constitute one way for computer systems to collect information about their—and our—shared environment, these systems are developing perceptions that far exceed our own. Much of what we do, how we live, how we interact with institutions, organizations, and one another takes place online, is readable telemetrically, and leaves clues about where we've been and where we're going. When you make an appointment and save it to the calendar application on your iPhone, when you leave your house and set a home alarm that dialogues directly with your city's police department, when you activate your phone's GPS, when you use your debit-procured Metrocard to access the subway and then use a radio frequency identification (RFID) enabled security tag to enter your office, you've created a trail that's transparent to anyone (or anything) with access to the servers and hard drives on which that data is stored. How big is that trail? Between checking your phone, using GPS, sending e-mail, tweets, and Facebook posts, and especially streaming movies and music, you create 1.8 million megabytes a year. It's enough to fill nine CD-ROMs every day. The device-ification of modern life in the developed world is the reason why more than 90 percent of all the data that exists was created in just the last three years.[2] Most of this is what's called metadata: bits of information that you create (or your devices make on your behalf) through your digital interactions. Only about 10 percent is ever stored permanently and very little of it affects you directly but all of it says something about you. And it's growing exponentially. There will be forty-four times as much digital information in 2020 (35 zettabytes) as there was in 2009 (8 zettabytes) according to the research group IDC.[3]

We think of each of these actions—the making of an appointment, the purchase of that fare through your subway fare card, the swiping of that RFID-enabled security badge—as separate ones of no real consequence to us, as big data. Think of that data instead as *sensory data*, as pinpricks that can be felt or sounds that can be heard like musical notes. The little actions, transactions, and exchanges of daily life do have a rhythm after all, and correspond to one another in a manner not unlike a melody. If you're like most people, your life has a certain routine: you leave for work at the same time each day; you shop at the same stores on your lunch hour; you take the same route home. Any tune composed of a repetitious sequence of notes becomes predictable. With sensors, geographic information systems, and geo-location-based apps, more of those notes become audible.

You've probably never heard this song. In the big data present, it's distant companies, market, and government forces that pick up the sound of our metadata. But this book isn't about the present. In the naked future the song is audible to everyone. The devices and digital services that we allow into our lives will make noticeable *to us* how predictable we really are.

The different ways we relate to the future publicly and personally will fundamentally change as a result of the fact that we will be making far more accurate and personal predictions. Huge areas of the future will be exposed. It will truly be a naked future.

The Future App

Throughout this book I refer to various hypothetical programs or apps like the one at the start of this introduction. These could be cloud-based programs we access on our smartphones, augmented-reality headsets, Microsoft brain implants (the blue screen of death would be literal in this sense), or any future platform. Although there are several apps such as Osito and Google Now that already use personal data to deliver customized predictions, most of the future-predicting apps in this book are made up. What they represent is the end point where telemetric data combine with processing to present an end user with a snapshot of the future. Though the future apps come from big data, just as we evolved from earlier humans, what they represent is something very different: an individual answer or solution to a unique, personal problem.

Predictability based on an abundance of personal data rises in almost direct inverse proportion to private data's remaining private. So how do we protect our privacy in the digital age?

In researching this book, I talked to people at Google, Stanford, MIT, Facebook, and Twitter; I hung out with hackers, entrepreneurs, scientists, cops, spies, and a billionaire or two. I was amazed by the promise of the telemetric age. I'm a future junkie. I get excited listening to smart people with world-changing ideas because if I didn't, I would be a pretty poor science journalist. But when I shared my experiences with friends, family, and colleagues and listened to their point of view, I realized that my reaction was not typical. Where I saw a thrilling and historic transformation in the world's oldest idea—the future—other people saw only Target, Facebook, Google, and the government using their data to surveil, track, and trick them. They were firmly planted in the big data present, in which it is us against them. They all had the same question: What can you do to prevent all of this from happening?

The threat of creeping techno-totalitarianism is real. But the realization of our worst fears is not the inevitable result of growing computational capability. Just as the costs of using big data have decreased for institutions, those costs will continue to trend downward as systems improve and as consumer services spring up in a field that is currently dominated by business-to-business players. The balance of power will shift—somewhat—in favor of individuals. Your phone may be from Apple; your carrier may be AT&T; your browser may be Google; but your data is yours *first* because you created it through your actions. Think of it not as a liability but as an asset you can take ownership of and use. In the naked future, your data will help you live much more healthily, realize more of your own goals in less time, avoid inconvenience and danger, and, as detailed in this book, learn about yourself and your own future in a way that no generation in human history ever thought possible. In fact, your data is your best defense against coercive, Target-like marketing and perhaps even against intrusive government practices. Your data is nothing less than a superpower waiting to be harnessed.

We still have choices to make. I'll discuss some of the forms those choices will take. But the worst possible move we as a society can make right now is demand that technological progress reverse itself. This is futile and shortsighted. We may be uncomfortable with the way companies, the NSA, and other groups use and abuse our information but that doesn't mean we will be producing less data anytime soon. As I mentioned earlier, according to the research group IDC there will be forty-four times as much digital information in 2020 as there was in 2009.[4] You have a clear choice: use your data or someone else will.

This is not a book about a change that is going to happen so much as a change that has already occurred but has yet to be acknowledged or fully felt. This is not a declaration of independence from corporate America, the government, or anything else. It's the record of our journey to this new place: the naked future.

CHAPTER 1
Namazu the Earth Shaker

THE date is April 12, 2011. I'm on a highway in the Japanese prefecture of Fukushima, home to a now infamous nuclear power plant that's in the process of melting down. I've just left the city of Ishinomaki where I was covering relief efforts that began following last month's earthquake and tsunami and I'm now heading back to Tokyo. In the car with me are two Japanese fishermen who speak no English, an Australian fireman named Simon, a British reporter stringing for a newspaper out of the Middle East, and a Japanese relief coordinator. Our route is taking us well within the eighty-kilometer "evacuation zone" that the U.S. government has advised its citizens to stay the hell out of. None of us have any illusions that it's safe to be here. For this reason, and because we're behind schedule, we're driving extremely fast.

Everyone on this road is driving fast.

Suddenly, a loud, sirenlike noise tears through the car's interior. Simon pulls his walkie-talkie from his Gore-Tex jacket. A bright red light cries out in distress at rhythmic intervals.

"Pull over," Simon commands. The driver applies the brakes,

not exactly slamming them but not gradually depressing them, either, and steers the car to the side of the road. Like a surreal piece of choreographed theater, every other car on the road also slows and banks.

A moment later, we feel the ground beneath us rise and fall. This is a 6.0 tremor, large enough that—had we been traveling at our previous speed of more than eighty miles per hour—we likely would have crashed. The fishermen, Simon, the car's other occupants, and I look around at one another. We share a silent acknowledgment that we have just barely avoided a terrible accident.

I'm alive today thanks in part to Japan's Earthquake Early Warning (EEW) system, a network of more than four thousand seismographic sensor stations.[1] These devices detect the low-level initial tremors called primary waves or P-waves that are released by seismic activity. An earthquake's P-wave telegraphs the size of the secondary wave or S-wave, the tremors that crash cities and bring the fury of the sea to shore. The system computes the signals as input and issues output, the feedback of which takes the form of Simon's phone going off.

The alert is issued automatically the second that the seismometer detects the signal and transfers it to headquarters.

Because earthquakes are a frequent occurrence in Japan, the alarm now goes off so often it has almost become background noise. In the moments before the 2011 earthquake hit, television broadcasts across the country were briefly interrupted by a crisp, telephonic ringing. A bright blue box appeared on every television screen showing the eastern coast of Japan and a large red *X* offshore depicting the earthquake's epicenter. In one of the eerier video clips that emerged from March 11, 2011, members of Japan's parliament can be seen debating a piece of legislation. Because they're accustomed to the signals they're slow to react to the warning at first. When they realize the size of the earthquake, they look nervously to the swinging chandeliers above them. The picture cuts to a flustered anchorman who warns of a possible tsunami off the coast of the prefecture of Miyagi.[2]

The Japanese have been applying creativity and resourcefulness to earthquake prediction for centuries. Historically, national myth held that earthquakes were caused by the movements of a giant catfish, or *namazu*, called the Earth Shaker. Though the idea seems ridiculous today, the Japanese took it very seriously at various points throughout their history. In 1592 the samurai warlord Toyotomi Hideyoshi issued what is perhaps the strangest building-code edict in history to the men constructing his castle in the Fushimi district of Kyoto: "Be sure to implement all catfish countermeasures."

In the later Edo period small catfish were awarded a reputation as earthquake predictors. Strange fish behavior was thought to be an indication that the giant *namazu* was on the prowl for mischief.

Today, the idea feels fanciful. Several centuries of steady scientific progress have taught us to look for concrete causal relationships in order to understand how one physical entity might influence another. We know that the earth's tectonic plates are affected neither by subterranean fish, nor the position of the constellation of Cassiopeia, nor the current level of God's wrath but by physical systems of enormous complexity and limited accessibility. Our understanding of the world through the lens of science suggests that P-waves indicate S-waves, but there exists no physical mechanism by which a catfish could know of an earthquake days in advance. Anecdotal evidence to the contrary proves only that humans have active imaginations, because catfish don't predict earthquakes.

Turns out, they almost do.

One of the key triggers of large seismic events is the buildup of pressure between rock formations in the earth's crust. This pressure also releases electrical activity and will do so days before large quake events. Loose "defect electrons" rise up through porous gaps in the earth's crust; they ionize when they meet the air. Under the *right* circumstances, this can cause subtle hydrogen peroxide increases in certain fault lines proximate to bodies of water, making such bodies just a bit toxic to very sensitive marine fauna.

British zoologist Rachel Grant observed this phenomenon firsthand when hundreds of toads fled a pond near L'Aquila, Italy, in the days just prior to an enormous 2010 earthquake. As Grant wrote in her paper that was published in the *Journal of Zoology*, "Our study is one of the first to document animal behavior before, during and after an earthquake. Our findings suggest that toads are able to detect pre-seismic cues such as the release of gases and charged particles, and use these as a form of earthquake early warning system."[3]

Catfish, like toads, have extremely sensitive skin. But unlike toads, they can't abandon a body of water that's becoming toxic. They can only thrash about or behave strangely, like the Earth Shaker.[4]

In his book *The Signal and the Noise: Why So Many Predictions Fail—But Some Don't*, statistician Nate Silver is rather hard on Grant. He suggests, though not explicitly, that she's reached an insupportable conclusion, as her paper seems to assert that the observed toad behavior is "evidence that they [toads] had predicted the earthquake." He describes her work as the sort of thing that "exhausts" real seismologists and notes dismissively, "Some of the stuff in academic journals is hard to distinguish from ancient Japanese folklore."[5]

Silver is certainly a talented statistician deserving of the celebrity that's been awarded him. He's right to point out that history is littered with failed attempts to predict earthquakes, often by observing strange animal behavior. He's also right to point out that statistical analysis of previous earthquakes is surely a far more useful signal than is toad behavior, at least for now.

But he's misstating Grant's intent. She's not suggesting that the toad behavior is "evidence that they predicted the earthquake." Neither the toads of L'Aquila, nor the catfish of Japan, nor even the EEW are actually predicting anything and Rachel Grant knows this perfectly well. These are feats not of prognostication but of detection. Grant and her colleagues acknowledge that testing the hypothesis outside a laboratory setting has thus far been impossible because they still don't know when and where an earthquake will strike. And neither do the toads. When they're in a pond with higher hydrogen peroxide levels they become uncomfortable and they leave. They are indifferent to earthquakes, to Nate Silver, and to the future.

It's humans who predict things.

As we attempt to make use of this abundance of telemetric data, we're going to make errors. One of the statistical traps Silver and other statisticians warn against is overfitting, or applying a specific solution to a general problem. In the case of earthquakes, this could mean watching toads rather than history because toad behavior lends itself to a very specific type of prediction method.

We are about to enter a golden age of overfitting, if such a thing can be said to exist. The sheer volume of data we now generate as individuals and institutions suggests that more people will be able to create more models with data points and observations that offer the false promise of certainty. We will model more and so we will make more errors, but an increase in modeling activity will not diminish the costs or consequences of those errors. Many small mistakes will feel extremely large particularly in the context of international stock and commodities markets. Overfitting also speaks to an impulsivity that's in our nature. We gravitate toward evidence, data, and facts that support a conclusion we've already reached or bolster the argument we're trying to make. Finally there's enough data to lend some support to virtually *any* argument, no matter how crazy. To overfit is human.

But the fact that electrical activity from pressure increases days before large seismic events is beyond dispute. It's exactly the sort of predictor that could reliably indicate an approaching disaster if only humanity could devise some cost-effective way to place millions of sensitive electron detectors deep beneath the earth's surface near fault lines. It's science fiction. But at one point, so was the idea of a sensor spiderweb that could detect P-waves.

We are turning our physical environment into a catfish.

A Global Nervous System Emerges

In 1988 a scientist at Xerox PARC named Mark D. Weiser put forward a novel vision for the future. Computer hardware, he said, would migrate from deskbound PCs to pads, boards, and "smart" systems that were part of the physical environment. The term Weiser gave this new sensing environment was "ubiquitous computing."

This vision for the future speaks a lot about the man who came up with it. Weiser was not a typical computer hardware genius. Take a look at his informal writings and the accounts of people who knew him and you will not find a man who loved gadgets and code for their own sake but someone motivated by a passion for *actual* experience, a sensualist, a devotee of skydiving, rock repelling, and lead drummer in a punk band called Severe Tire Damage. Through ubiquitous computing he imagined a future in which humans interacted with computers on an unconscious level, through regular activity; a future in which computers served to remove annoyances and answer questions like "Where are the car keys?" "Can I get a parking place?" and "Is that shirt I saw last week at Macy's still on the rack?"[6] while keeping us connected to what we care about. Computers weren't supposed to get in our way, or be constantly in our hands, or be connected to our ears through shiny white earplugs, or demand that we answer their every chirp and bell. As they became better they were supposed to become more numerous but also disappear into the background.

A decade after his death, it's the "ubiquitous" portion of Weiser's ubiquitous computing vision that's becoming reality for most of us. The total number of devices connected to the Internet first exceeded the size of the global human population in 2008 or so, according to Cisco, and is growing far faster.

Cisco forecasts that there will be 50 billion machine-to-machine devices in existence by 2020, up from 13 billion in 2013. Today, we call ubiquitous computing by another name: the Internet of Things.

For large institutional or corporate consumers of information, the spread of sensors and computer hardware across the physical environment amounts to better inventory tracking and customer targeting, which will help bottom lines. The Internet of Things can be found most immediately in the RFID tags that have made their way onto everything from enormous inventory palettes to the clothing labels that Swiss textile company TexTrace[7] sews into American Apparel clothing to track shipments. Most RFID tags that we encounter today are small squares of paper, plastic, or glass containing a microchip and an antenna at a cost of about twenty cents. The microchip holds information about the product (or thing the RFID is connected to). The antenna allows an RFID reader to access data on the chip via a unique radio signal. Unlike a simple printed bar or quick response (QR) code, the RFID tag doesn't have to be directly under the reader to work. The reader need only be close by. This allows retailers to monitor the inventory in their store in something like real time. Some futurists have suggested that RFID could one day render the checkout station obsolete. In this future, when you saw a product that you wanted you would simply pluck it from the shelf and—so long as you had a user account or were identifiable to that store—walk out the door. The product's RFID tag would tell the retailer the product had been purchased and your account would be debited. Sound far-fetched? Millions of Americans today *buy* access to toll roads through the dashboard-mounted RFID tags that are part of the E-ZPass system. The act of purchasing takes the form of a simple deceleration and a brief exchange of data between the RFID tag's antenna and the tollbooth's reader. And RFID is just one of the many smart or sensing tags and microchips that are making their way into our physical environment at rapidly decreasing cost.

For patients and graying baby boomers, the Internet of Things is ushering in a revolution in real-time medical care. It is alive inside the chest of Carol Kasyjanski, a woman who in 2009 became the second human being to receive a Bluetooth-enabled pacemaker that allows her heart to dialogue directly with her doctor.[8] The first was former U.S. vice president Dick Cheney, who received one in 2007, but never activated the device's broadcasting capability for fear of hackers.

The military uses the Internet of Things to do more with less. In Afghanistan it takes the form of the fifteen hundred "unattended ground sensors" that the U.S. Army is leaving littered across the Afghan countryside as the U.S. mission there winds down. These sensors, which are intended to pick up human movement, are intended to allow the Pentagon to eavesdrop on the countryside and detect how Afghans (or Pakistanis) are moving over their country.

It is, quite simply, all of the computerized sensory information that can be gathered and transmitted in real time about what is happening right now. When this happens to machines we call this big data. When it happens to us we call it sensing.

In many ways, this expanding, computer-connected environment is inconspicuous (as Weiser intended). The presence of sensors able to detect ammonia, a common component of explosive material, in the New York City subway is not something I devote thought to when I'm taking the downtown 6 train; I'm just glad it's there.[9]

The Internet of Things is not a far-off dream; it's here. We've been accepting the presence of more sensors in our environment for decades now. It's impossible to argue against the usefulness of Japan's EEW, or radon detection devices in subterranean structures, or home security systems that sense when a door is being opened and alert the police and homeowner. The average 777 has so many sensors on board that a three-hour flight can generate a terabyte of data. Twenty flights generate the data equivalent of every piece of text in the Library of Congress.

For the owners of the copper wires, the fiber-optic cables, the cell phone towers, and the servers on which the Internet runs, the growth of the Internet of Things means massive future profits. The firm Gartner has predicted that the global market for "contextually aware computing" will exceed $96 billion per year by 2015. It's no wonder such companies as Cisco, IBM, and Verizon spend millions of dollars in ad, marketing, and grant campaigns to persuade the world that a "smarter" planet is so very good for everyone. And it is, in many ways. But first and foremost, a smarter planet is good for them.

Importantly, the Internet of Things is not solely the product of companies and governments. It's become a homegrown phenomenon as much as a big telecom money machine, and it's empowering regular people in some very surprising ways.

The Internet of Things, Three Vignettes

On March 11, 2011, engineer Seigo Ishino was at his office in the city of Kawasaki near Tokyo when the EEW system sounded. Like any rational person caught in a massive tremor, he crawled under his desk until the quake passed. He emerged a few minutes later unscathed but, as a result of the seismic event that had just occurred, his life was now far more complicated than it had been just that morning. News of technical problems at the Fukushima Daiichi nuclear plant spread quickly in those early hours through Japan and then around the globe. Tokyo was close enough to Fukushima (about 160 miles) that the meltdown posed a serious concern particularly for children and pregnant women, as radiation is most harmful to babies and kids. Seigo's wife was eight months pregnant at the time. He was faced with some hard choices. Was the level of radioactive cesium and iodine spewing out of the plant dangerous enough to compel him to relocate his family farther south? If so, he needed to act quickly to get a train ticket, as the price was rapidly ascending. There was also the question of where they would stay, and how he would earn money because he would effectively be abdicating his duties at his present job and had no job prospects outside of Tokyo. Was the danger significant enough to warrant an evacuation from Japan? The many thousands of foreigners who were attempting to leave that week were also driving up the cost of airfare and there were questions about how to obtain an exit visa. Alternatively, was it safe to stay where he was? What about the food and the water supply? He needed more information.

The Kan administration's press secretary, Yukio Edano, began giving regular press conferences, clad in a bizarre blue jumpsuit, to inform the public that radiation levels were not dangerous and that the situation at the troubled plant was under control. The official messaging took a turn for the ridiculous on March 12 when the Kan administration assured the public that the pressure levels at the reactor had stabilized only to then admit, a few hours later, that a massive buildup of pressure had blown the walls off the reactor building. Yukio Edano again took to the podium to steadfastly affirm that the situation was improving as the reactor in the picture behind him smoked and fumed.

Seigo elected to stay but, like millions of other Japanese, he no longer trusted the official story that was coming from the government and from TEPCO, the corporate entity that operated the plant, both of which he regards as "most untruthful."

Seigo was a member of an international group of community designers, engineers, hackers, and hobbyists who built sensors and installed them in buildings and other aspects of the built environment to monitor energy use. The community was centered around a platform called Pachube (now Cosm), which allows users with sensor data to share it in real time on the group's site. Not long after the news of the meltdown spread throughout Japan, thousands of people across the country were tweeting Geiger counter data and hundreds of Pachube users were streaming their data directly to the Pachube site.

Seigo began work on a smartphone app (for Android) that combined Google Maps, real-time information about radiation levels, and publicly available data about wind currents. The resulting Winds of Fukushima app worked as a sort of living map that provided constant information not just on where radiation existed but also *where it was going*, in the form of bright blue arrows.

Of particular concern to Seigo was the safety of food and drinking water. The Winds of Fukushima app confirmed that radiation was spreading far wider than the government was indicating in news reports. Seigo began to buy his food from the south side of Japan and drink water imported from the United States.

Winds of Fukushima is hardly a technological miracle. It takes a very conventional stream of data (current wind direction), combines it with a second data stream (real-time readings of radiation), and makes this new, combined data available in a format that the public can easily find and use: a Google map. Its most revolutionary aspect is how quickly it emerged in the wake of the disaster. A decade ago, the task of coordinating among hundreds of Geiger counter–armed volunteers, building a platform for all of them to stream data, and finding a vendor willing to sell the software internationally was neither cheap nor easy. Thanks to communities of interconnected amateur techies, open APIs like Google Maps, and direct-to-market software vending platforms like the Android app stores, Seigo was able to build and publish Winds of Fukushima from a small Yokohama apartment in virtually no time at all. The app went live in the Android store about six weeks after the initial quake but it actually took Seigo only a few days to create it (though he admits he barely slept).

Pachube was started by an architect named Usman Haque who wanted to build a sensing feature into his building designs so that, years after construction, he and his fellow designers could log on to Pachube and get a sense of how the buildings were being used. He wanted to let the occupants, too, reconfigure their living environments around their actual use patterns, their living data. Today, the Cosm system that acquired Pachube allows developers to build apps, programs, and immediately derive insights off massive amounts of data coming from a suddenly awake world.

"Everyone gets insight into the environment around them, data contributors get applications that are directly relevant to their immediate environment, and application developers get access to a marketplace for their software," Pachube evangelist Ed Borden remarked in a blog post.

A world that senses its occupants and shares that information may be one where people become much smarter about how they live. It's also a world where information that is accessible only to government suddenly becomes available to hackers and activists. Depending on the content of that information, and the method you go about obtaining it, a simple civic act such as trying to fix your local sewer system can look provocative to the local authority whose power you just usurped, as another Pachube user named Leif Percifield discovered in New York.

Seeing the Hot Water Before It Hits the River

The date is April 18, 2012. Leif Percifield, a few of his friends, and I are in canoes in Brooklyn's famous Gowanus Canal. A shy drizzle rains down on us as we paddle out over rusted bicycles, tin cans, and other bits of metal and plastic that have imbedded themselves in this canal bed. The air smells slightly of sewage, which is why we're here.

We reach our destination, the portion of the canal that meets Bond and Fourth streets. Leif secures a shoe-box-size plastic container with a solar panel atop it to a mooring above the combined sewer overflow pipe, or CSO. Two long wires extend from the device; these are tipped at the end by a small sensor. Leif made the box, which is a prototype, the day before using off-the-shelf components (an Arduino motherboard) and parts he created himself with a printable circuit-board machine at Parsons School of Design. Leif plunges his hands into the water, elbow deep, to affix the sensor tip as close as possible to the pipe. When he's done, he does a cursory clean of his hands and checks his iPhone.

"It works!" he says. The sunken sensor is now broadcasting the temperature and conductivity of the water. Hotter water, and water with more electricity conducting minerals, are sure signs of sewage runoff.

Just about everyone in New York takes for granted a few key facts about the Gowanus Canal. The most important of these is that it's beyond fixing. Not only does sewage water run into the canal when it rains but the water is laden with decades' worth of heavy toxic metals, which has earned its designation as a Superfund site, one of the most poisonous environments in the United States. The U.S. Clean Water Act says the city of New York is supposed to clean this place up, remove the metals, and keep sewage from spilling into its waters. But before that can happen New York City and the U.S. Army Corps of Engineers must conduct a feasibility study, which neither New York City, the EPA, nor the army are in any hurry to complete.

"The numbers that we have say that three hundred million gallons of sewage go into the Gowanus Canal a year," says Leif. He adds that the numbers are based on computer models and he believes them to be flawed. Members of the community have accused the New York Department of Energy of tweaking the data in order to put off the costly work of fixing the storm runoff problem.

Leif's goal is to get people in the city to participate in rehabilitating the canal. That's not easy. But if he can map where flows are bigger or smaller, he thinks he can put together a more accurate assessment of what's going on and essentially predict how dirty the water will be on any given day as a result of environmental factors. This information is of no real use to one person but a community can edit their water usage, their showering and flushing, based on the sewage water level. The name of Leif's blog says it all: *Don't Flush Me.*

You would expect the city would appreciate Leif's efforts to better monitor the sewer system. But his relationship with local New York City authorities quickly became rocky. His previous project literally got him in a lot of hot water: he actually went into the city sewer system to fit it with a network of sensors.

"The air is not pleasant," he says of the New York underground. "But I was thinking it would be putrid. Instead it was more acrid. And it was incredibly hot, twenty-five degrees hotter underground than aboveground. People use hot water, you know, and hot things come out of your body."

The sensors he attempted to install were supposed to read the water level and a fast rise was a good indication of coming overflow. The experiment didn't pan out. The sensors didn't stay in place and the Bluetooth signal inside the sewer was too weak. The data was trapped. Had the sensor system functioned, he would have been the second person in history to be able to predict when the sewers were going to overflow. The first was Cynthia Rudin, an MIT researcher who figured it out with a statistical formula.

Leif's project was simple, commonsense infrastructure stewardship. But when he posted a few pictures of his adventure online he immediately got a call from New York's city officials. They ordered him downtown and made it clear that he was "in trouble" for what he had done. He was told to cease his activities. Leif believes this is because of how he was able to show how easy it is to get into the New York City sewage system.

Leif has grown better at working with the New York City Department of Environmental Protection but his experience reveals how complicated our relationship with authority becomes in this interconnected era. The program Leif put in place all by himself was very similar to ones in place in Maryland and Washington, D.C., to manage sewage run off in the Chesapeake Bay, but the latter are managed by local authorities with little citizen input so they're less controversial (and arguably rather ineffective). Everyone can agree, at least publicly, that fixing sewage backup should be a top priority. But when citizens armed with sensor boards suddenly start outflanking government on government's own turf, tensions can rise.

Most of us grew up in an environment where we comfortably assumed that local government always had more information than we did about what was going on within our city, certainly the best data on the state of infrastructure. We also instinctively trust local government as the provider of information during an emergency, even when it's an emergency in which we're directly involved. See a fire? Call 911 and ask for services, wait for someone to come to where you are and tell you what's happening. This is an inefficient way to collect and distribute information during a time of crisis.

The Internet of Things is ushering in a new era of proactive citizenry. It's an era where much of the most important information during a fire, a flood, a citywide disaster doesn't come from government but from you and your suddenly empowered neighbors, people like Gordon Jones.

Seeing the Fire Before You Are in It

In the summer of 2007 Gordon Jones was living in Charleston, South Carolina. A fire broke out at a nearby furniture store, killing nine firefighters, the largest number of firemen to die on duty since the 9/11 terrorist attack. An enormous memorial followed. Emergency workers from around the country came out to Charleston as the facts of the incident were reported on the news in rounds, like a funerary dirge. The public safety workers succeeded in pulling out several survivors from the blaze before the roof caved in on them.

Jones was working at the time for Global Emergency Resources (GER), a company that markets a software tool for monitoring ambulances and hospitals during emergencies. Watching the local coverage of the memorial service for the firefighters, he realized that the technology he was developing could have saved lives: "I said to myself, *What if somebody, one of the people trapped inside the store, had a smartphone to broadcast what the scene looked like?* That might have made a difference."

It sounded like a worthwhile and potentially profitable start-up. Jones founded a company and shortly after announced the launch of the Guardian Watch app. Guardian Watch enables anyone with a cell phone to live-stream video and pictures of an event directly to emergency personnel. This may not sound that significant but think of an alarm system as nothing more than an information distribution network. Some alarm systems are better than others. Guardian Watch enables thousands of people to provide streaming visual data about a situation at an information transfer rate of hundreds of thousands of bytes per second, the average upload speed of a 4G or higher phone. Guardian Watch was the first iPhone app to take advantage of the smartphone's full capabilities to give emergency workers a visual and auditory sense of what may be ahead of them.

"A picture is worth a thousand words and a video is worth a thousand pictures," says Jones. This statement, though perhaps a bit corny, encapsulates why Guardian Watch really is a clear improvement over traditional emergency response systems. It delivers information that's user-specific, varies depending on context, and moves at a speed and scale that make sense for emergencies—namely more and faster.

A decade ago, increasing the scale of information collection and distribution to the point where it would have made a difference to one of those Charleston firefighters would have been a daunting technological challenge. Today, the tools, platform, and infrastructure already exist and have been widely distributed. You're carrying all of this around in your pocket.

The single biggest driver of the Internet of Things is the smartphone, that always-on, GPS-enabled sensor that more than 64 percent of the U.S. population carries around with them. We know that smartphones today make it easier to find restaurants, share experiences as they occur, shop, and study. Mobile technology makes data creation and curation possible anywhere, which means we're creating and curating much more of that data more of the time.

Guardian Watch already faces competition from other groups looking to leverage the information gathering and broadcasting technology of smartphones. A Silicon Valley–based start-up called CiviGuard takes the idea a step further. The platform integrates streams from Twitter, Facebook, and local emergency channels and presents the user with a "networked window" of an emergency situation playing out in real time. It gives geo-tagged advice that's specific to an individual user based on a variety of variables, the most important of which is location. What that means is this: depending on the situation, a user may be told to stay where she is while a different user may be told stay away from that area. Most important, CiviGuard includes a scenario function to allow users to conduct virtual emergency simulations.

Imagine you're in Manhattan and there's just been a terrorist attack. Want to know which streets are most likely to become blocked when the news spreads? How your company's supply chain will be disrupted? Where to find food and water while they're still available on store shelves? CiviGuard will tell you and will do so based on a rapidly updating understanding of what's going on around the city. And should CiviGuard not pan out, the Environmental Systems Resources Institute (Esri) can also build you a custom geographic information system that does all of the above, and can integrate it with population density, water tables, jurisdiction, and hundreds of other maps.

The same real-time broadcasting capability that will allow me to better navigate my way out of a disaster can be used for other purposes as well. The mapping of human behavior promises enormous benefits, but it also speaks to a future where invisibility and anonymity are no longer the default setting for life.

The Internet of Things is also the intersection camera that snaps a picture of my license when I try to beat the yellow light. It's the smart electricity meter that California's Pacific Gas and Electric Company now insists its customers use, allowing the utility to optimize energy delivery but also to better track individual energy use. If you're a PG&E customer, the Internet of Things is the reason why your energy company can infer when you're home and when you're not based on when and how you use certain devices.[10]

In our rush to overlay data collection devices across the physical environment, we overlooked the fact that the same devices we use to perceive our environment can just as easily be turned on us.

We will be seen. We will be tagged. It's happening.

Checked In. Your iPhone Knows Where You're Going

Want to know how many people with smartphones are in terminal 4 of New York's JFK Airport, standing on line to get tickets to *The Daily Show*, browsing the shoe store down the street? A company called Navizon sells a device that can track every phone using Wi-Fi within a given area. Just plug this device into a nearby wall outlet to monitor that action in real time. Because we know that more than 60 percent of the U.S. population now owns a smartphone, a couple of months' worth of data will tell you how many people are likely to be in any area that you're surveilling on any given day and time of the week. Leave the device plugged in for a few decades and you'll have a *reasonable* estimate for how many people will be at a specific place, at a specific time, on a specific day of the year. This is the sort of information that big phone companies like Verizon and AT&T have at their fingertips. When you walk around with your cell phone on, you give these companies data about your location. AT&T and Verizon then strip that data of identifying information and sell it to city planners, commercial interests, and others. Verizon even claims the ability to build a demographic profile of people gathered together in a specific place for a specific thing, such as in a stadium for a rock concert or a sporting event.[11]

Navizon puts the same sort of capability in the hands of individuals with small budgets but larger time horizons. Navizon CEO and founder Cyril Houri is marketing the device as a way for entrepreneurs to do location planning. There are some limitations. Because the device measures Wi-Fi from smartphones, it's also biased toward younger adults (18–25) who are—not surprisingly—more likely to own a smartphone than are people over age sixty-five. High-income earners also show up more often than low-income earners. But the current profile of smartphone users is not the *future* profile.

Navizon's analytics system won't disclose the names of specific people whom the device picks (unless those people opt in to the Navizon buddy network) but the system can recognize individual phones. It has to, in order to count them. If someone follows roughly the same pattern every day, hitting work, the store (or the bar), then home in the same time window, the difference between tagging the phone and tagging the person effectively disappears. MIT researchers Yves-Alexandre de Montjoye, César A. Hidalgo, Michel Verleysen, and Vincent Blondel from the Université catholique de Louvain took a big data set of anonymized GPS and cell phone records for 1.5 million people, the sort of stripped-down location data that Verizon and AT&T sell to corporate partners to figure out the types of people who can be found at specific locations at particular times of day. The data consisted of records of particular phones checking in with particular cell antennas. What the researchers found was that for 95 percent of the subjects, just four location data points were enough to link the mobile data to a unique person.[12]

A growing percentage of smartphone users voluntarily surrender data about themselves wherever they use geo-social apps. Facebook, Twitter, and Google+ all have "check in" features that broadcast your location to people in your network. Other, more creative services will facilitate specific interactions based on what you're looking to do wherever you happen to be.

An app called Sonar will identify the VIPs in the room; Banjo will tell you the names of nearby Twitter, Facebook, and Instagram users; a service called Grindr, launched back in 2009, will pinpoint the location of the nearest gay man who may be interested in a relationship—of either the long- or short-term variety.

To the smartphone-suspicious, these services seem to be more trouble than they're worth. What's the value of knowing the Twitter handle of the person at the next table in a restaurant, when, at best, such an app just detracts from the authentic experience of real life? At worst, it's giving away personal info to strangers.

However, to a growing number of smartphone owners, check-ins and geo-social Web apps like Foursquare are an integral aspect of smartphone ownership. More than 18 percent of smartphone owners use some sort of geo-social service (as of February 2012), a number up 33 percent in one year, with heaviest use concentrated among the young. Importantly, more than 70 percent of smartphone owners use *some* sort of location-based service on the phone, even if it is just the GPS.[13]

These apps change the way users perceive and interact with their environment as well as the way actors in that environment interact with them. Geo-social apps work to raise the net awareness level in any neighborhood or room. Today, most of this added social intelligence is of limited value at best. But the situation is evolving rapidly.

The rising popularity of these apps, which is closely connected to smartphone adoption in general, promises a big change in our expectations of privacy. There's inevitability to this. As more people buy smartphones, more people use them the way the devices were designed to be used, with geo-social and location-aware apps. Wearable computing, if it eventually replaces what we know today as cell phones, will further enable this trend. We want to know more about the environment we're in, what people on Yelp, contributors on Wikipedia, and friends on Facebook have to say about the place where we've arrived. This is what futurist Jamais Cascio calls augmented reality, and what the U.S. Department of Defense (DOD) calls situational awareness. It's also human nature. As our friends, neighbors, nieces, nephews, sons, and daughters submit to the impulse to download an app that uses location information, the opt-out strategy becomes less effective for the rest of us, even those of us who consider ourselves extremely privacy aware.

We leak data through our friends.

One of the better-known examples of the accidental surrender of personal information via smartphone—what hacker, author, and astrophysicist Alasdair Allan has dubbed data leakage—involves an app called Path, which was billed as a smarter, leaner, more mobile-friendly answer to Facebook. Started by Facebook alum Dave Morin, the service was launched as a way for users to digitally document comings and goings in the world. This was your path. The service worked a lot like Facebook except that users were limited to 150 friends, based on the theory that 150 is the maximum amount of useful acquaintances that a person is capable of maintaining. These people would receive the premium subscription to your ongoing life story. Path received angel-investor funding from the likes of Ashton Kutcher and after tweaking the service a bit, it went from 10,000 users to 300,000 in less than a month.[14,15] The service today has more than 10 million users.

Path was a hit because it seemed to provide the sort of intimate, authentic, and secure sharing experience that Facebook couldn't offer once users had to have different privacy settings for bosses, English teachers, mothers-in-law, et cetera. The sharing and posting on Path felt intuitive. Turns out it was a bit too intuitive.

Before long, a Singapore-based developer named Arun Thampi discovered that the ease of interfacing came at a high cost. Thampi was playing around with the code when he discovered something unusual. "It all started innocently enough," he wrote on his blog. "I was thinking of implementing a Path Mac OS X app as part of our regularly scheduled hackathon . . . I started to observe the various API calls made to Path's servers from the iPhone app . . . I observed a POST request to https://api.path.com/3/contacts/add. Upon inspecting closer, I noticed that my entire address book (including full names, emails and phone numbers) was being sent as a plist to Path. Now I don't remember having given permission to Path to access my address book and send its contents to its servers, so I created a completely new 'Path' and repeated the experiment and I got the same result—my address book was in Path's hands."[16]

The company was holding detailed information on the friends, families, coworkers, and contacts for all three hundred thousand or so of its users, a list that potentially included tens of millions of people. They quickly issued an apology and a software update. But, in many ways, the damage was already done.

Allan has called this the inevitable result of the increasing market for mobile software among people who don't understand—and have no desire to learn—how their most cherished devices work. We want our apps to know us, to present customized answers to our problems and questions, but we don't care how they arrive at those solutions until there's a problem.

Most people who download Instagram, Twitter, or Facebook to their phone already understand, at least in part, that they're risking their personal private information in doing so. But they probably wouldn't elect to give their *grandparents'* contact information and other personal details to some strange company. Given that more than 9 percent of the entire U.S. population is part of a geo-social network (as calculated from the fact that 18 percent of smartphone owners are part of a geo-social network, and well more than 50 percent of the population owns a smartphone), further incidents of data leakage will affect the U.S. population well beyond the smartphone-owning community.

We presume that our personal data is compromised only when we *choose* to take a certain risky action. Maybe some people find amusement in these silly networks and don't mind giving away their information to strangers, but that shouldn't have any bearing on *me,* goes this line of thinking. But our friends and loved ones create data about life and that data includes us, whether we wish to be tagged or not. This is why we are using the wrong set of words to explain this phenomenon; we think of data leakage as an act of theft but we need to understand it as a contagion event. If you know someone who geo-tags their tweets, Facebook posts, or Instagram photos, you've already been infected.

Telemetry, Simulation, and Bayes

Once these signals are sensed, they must be processed if they are to form the basis of a useful prediction. But predictions—like the future itself—spring from the brain. The challenge is getting computers, programs, and systems to make predictions on the basis of continuously sensed information, on the basis of what's happening now in (sort of) the same way that the brain does. This is an entirely recent problem related to the rise of continuous data streams and all the artifacts of modern information overload. But the mathematical formula to tackle it has actually been around for centuries and can be utilized as easily by a college undergrad as by a roomful of scientists.

Researchers use plenty of statistical methods, and mathematical tricks can be employed, in isolation or combination, to turn data into a prediction. But the one method that allows you to make new predictions and update old predictions on the basis of new information is named after its founder, Thomas Bayes. The theorem in its simplest form is:

$$P(A \mid X) = \frac{p(X \mid A)\, p(A)}{p(X)}$$

In the above, *P* is probability, *A* is the outcome we are trying to predict, and *X* is some condition that could affect *P*. The theory solves for *A* given (|) *X*. The value you award *P* when you begin is sometimes called the "prior"; the value you award *P* after you've run the formula is called the "posterior."

Undeniably, compared with other statistical methods Bayes won't always give you the most *accurate* answer based on the data that you're looking at. But it does give you a fairly honest answer. A large gap (in value) between the prior and the posterior suggests a small degree of confidence.

Celebrated artificial intelligence (AI) luminary and statistician Judea Pearl describes the process as follows: the Bayesian interpretation of probability is one in which we "encode degrees of belief about events in the world, and data are used to strengthen, update or weaken those beliefs."[17]

Compared with many other statistical methods such as traditional linear regression, Bayes is one of the most like the brain. Predictions of probability combine past experience with sensed input to create a (somewhat) moving picture of the future.

What's important to understand is that although Thomas Bayes's formula wasn't published until 1764, about three years after his death, it's only in the last couple of decades that Bayes has come to be seen as the essential lens through which to understand probability in a wide number of contexts. The Bayesian formula plays a critical role in statistical research methods having to deal with computer and AI problems but also the simple questions of quantifying what may happen.

When I asked the researchers in this book why they found

Bayes more useful than other statistical methods for their work, the most common response I received was that Bayesian inference allows you to update a probability assumption—the degree of faith you have in a particular outcome—on the basis of new information, new details, and new facts, and to do so very quickly. If the interconnected age has taught us nothing else, it is that there will *always* be new facts. Bayes lets you speedily move closer to a better answer on the basis of new information.

Here's an example. Let's say it's Tuesday and you are scheduled to meet your therapist. Your therapist has never missed a Tuesday appointment so you hypothesize that the probability of her showing up is 100 percent. The P for A or $P(A) = 1$. This is your *prior* belief. Obviously, it's terrible. There is never a 100 percent chance that someone will show up to work. Now, let's say you get some new information, that your therapist has just left from a previous appointment and she is three miles away, on foot. How would you go about adjusting your belief to more accurately reflect the probability that your therapist will make it to your appointment on time?

Let's say you find some new data, that the average walking speed is 3.1 miles per hour. Given time and distance you can compute that your therapist will surely be late. But you must compute this in light of the prior value; your therapist is *never* late. You now know the chances of your therapist being late for this appointment are lower than they would be for a regular person but the possibility of her being late for your appointment, in spite of what you understand to be the lessons of all history, have grown significantly. Now you discover even more information: according to reviews of your therapist's practice on Yelp, she's actually late to her appointments about half the time. You can recompute the probability of your therapist's getting to the appointment on time over and over, every time you get some new tidbit that reveals reality more clearly in all its inconvenience. What is making the future more transparent is the exponentially growing number of tidbits we have to work with. Bayes lets us manage that growth.

Imagine next that you have an enormous amount of telemetrically gathered information to update your prior assumption. You can actually track your therapist moving toward you in real time through her Nike+ profile. You can read the wind currents meeting her via Cosm's feed off a nearby wind sensor. You can measure her heart rate and hundreds of other signals that might further refine your understanding of where she is going to be, relative to you, in the next few minutes. Let's say you also have access to an enormous supercomputer capable of running thousands of simulations a minute, enabling you to weigh and average each new variable and piece of information more accurately. The influence of your first hilarious off-the-mark prior assumption about your therapist's perfect punctuality is, through this process, dissolved down to nothing.

This is the promise of *sensed* data, of telemetrics combined with easy-to-update statistical tools such as Bayes.

Finding You

In March 2010, Adam Sadilek, a young Czech-born researcher from the University of Rochester, set out with some colleagues to see how accurately they could predict the location of someone who had turned off his or her GPS, who wasn't geo-tagging tweets or posts, who was in effect going incognito. Sadilek and his team sampled the tweets of more than 1.2 million individuals across New York City and Los Angeles (America's chirpiest cities). After a month, the team had more than 26 million individual messages with which to work; 7.6 million of those tweets were geo-tagged.

They trained an algorithm using Bayesian machine learning to explore the potential patterns among the Tweeters. The idea was to uncover the conversations between the users, contextualize what conversations were taking place across the New York and Los Angeles landscapes, and see if they could use that information to discover information about people who were friends with the geo-taggers but who weren't themselves geo-tagging.

Turns out that your friends' geo-tagged tweets provide a great indication of where you've been, even if you weren't in that place with that friend. Because you, like most people, are probably a creature of habit, where you've been is an excellent indicator of where you're going.

Let's say Sadilek's system has no "historical information" on you. You don't geo-tag tweets; you keep your phone's GPS setting off; you are invisible, a covert operative. But in order to maintain your cover, you established a Twitter account using a dummy e-mail address. Let's also say you've got two friends on Twitter. They're real friends, people you talk to about events in real life and with whom you relate in the real world. You see them in class, at clubs, in line at the post office. Like a lot of other people, these two friends do geo-tag their tweets. Sadilek's system can predict *your* location at any moment (down to 328 feet and within a twenty-minute frame) with 47 percent accuracy. That means he's got a 50 percent chance of catching you at any given moment.[18]

I know, I know, you did everything right. You were a careful steward of your privacy. It's not fair that a twenty-five-year-old PhD grad from Czechoslovakia should be able to find out so much about you so effortlessly. It was your friends who gave you away without even realizing it. Now your not-so-secret-agent career is over.

I went to meet Sadilek at an AI conference. Sitting in the executive lounge on the top floor of the Toronto Sheraton, we overlooked downtown and saw people parking their bicycles, waiting for buses, talking on phones, walking with heads pointed toward shoes, white iPod cords dangling from their ears, people coming and going from little secret rendezvous that every one of them presumed were unknowable to the outside world. We talked a bit about human predictability.

"Somehow, growing up as a teenager, I always was sort of put off by how predictable people are. I never liked that. I liked people that were random."

Since entering the field of machine learning, Sadilek has come face-to-face with a hard truth. Human behavior is far more predictable than anyone ever predicted; surprisingly predictable you may even say. One experiment in particular proved this in a way that astounded even Sadilek.

The year was 2011 and he was about to start an internship at Microsoft with researcher John Krumm. In his years of working at Microsoft, at a time when the company was at its most ambitious and adventurous, Krumm was able to amass a rather unique data set. He set out to make a sort of living map of human mobility the way zoologists and biologists track the movement of bears or birds or lions; but because Microsoft was so flush with cash at the time, Krumm paid several hundred test subjects to carry GPS trackers around with them wherever they went, which broadcasted the wearers' physical location every couple of seconds. Some people carried the trackers in their pockets and some had the tracker installed on the dashboard of their cars. Microsoft was considering a lot of potential uses for this data, from helping cities better understand traffic patterns to developing a new line of smart thermostats that could predict when customers were on their way home and accordingly turn on the heat. Another potential use was an intelligent calendar to be used in conjunction with Outlook (the default e-mail provider that comes with Windows), which could forecast your potential availability for appointments into the future. Krumm watched the trackers and the people to which they were connected sail through life for more than six years. Altogether, his seven hundred–plus subjects provided more than ninety years' worth of data on human mobility.

He presented the data set to Sadilek and they applied an algebraic technique called eigendecomposition to it. Decomposition in this sense simply means reducing a lot of numbers to a single value that's in some way characteristic of the whole. Eigen is derived from the German word for "self." Through eigendecomposition Sadilek and Krumm were able to create a model that could predict a subject's location with higher than 80 percent accuracy up to eighty weeks in advance.[19]

Put another way, based on information stored in your phone, Sadilek and Krumm's model can predict where you will be—down to the hour and within a square block—*one year and a half from right now.*

Granted, Krumm and Sadilek's data set isn't a typical one. Most of us don't share geo-location information as frequently as did the folks Krumm put on the payroll. At least not yet. And most of us bounce between home, work, or school and back pretty regularly. In fact, if you know where someone usually is on a Monday at 10 A.M. you can infer their location on any given Monday at 10 A.M. fairly well, but it's still just a guess based on two data points. The magic of Sadilek and Krumm's Far Out model, as they named it, is that it factors in the occasional random detour—the flat tire, the unexpected work junket, or the sick day—without making those outlier events more significant than they are, without overfitting.

A flat tire on a Monday at 10 A.M. isn't actually random, according to the strict definition of the word. We just don't yet know how to model it. A certain type of person, someone who wears his tires thin without replacing them, someone who drives through an area with lots of hazards, et cetera, is more likely to suffer a flat every few months than is someone who doesn't take her car out as often, or to the same places, or who replaces her tires religiously. Sadilek's system doesn't explain why some people have more flats, but it does find some people are more prone to these anomalies than others. When you have a data set with enough points, even outliers can reveal a pattern.

I asked Sadilek about how people respond to his work on the Far Out model when he tells them about it. Researchers, by and large, are intrigued and appreciate it. Folks outside the field, many of whom carry a GPS tracker in their pocket without ever realizing it, have a different reaction.

"A small amount of people always worry about the privacy implications of this," he answered the way a doctor may discuss the unfortunate symptoms of a chronic but medically interesting condition, as in, *Of course you will experience night terrors and eye bleeding; these are now just a part of your life.* The prescription: take this insight and modulate your behavior. "Now that you know, looking at our papers, what can be done when you tweet about your stuff," he told me. But it was clear he didn't actually *want* people to tweet less, to become less predictable. He just wanted them to feel differently about it, to be as excited about this discovery as he was when he made it. This speaks to an important point: the idea of someone else, a government body, a company, a stalker, knowing with more than 80 percent certainty where you will be a year and a half into your future sounds like a scenario from dystopian science fiction. Yet the ability to predict *your own* future location that far in advance qualifies as a superpower. Sadilek's breakthrough brings that power closer to reality for more people.

Here's an easy step you can take if your location predictability level worries you: simply turn off the geo-location feature on your smartphone unless you really need it, and certainly limit the number of apps that have access to your location data (then watch your battery life magically regain its youth). But make no mistake: while you can turn down the signal that you're sending out, that doesn't actually make you less predictable; it just makes your predictability level harder to detect. Your future is still naked even if no one has noticed yet.

As we watched the people below us from the Toronto Sheraton, Sadilek put it slightly differently: "Everybody needs a schedule that they can fixate on, otherwise they get really unhappy. If you live this random life, you always are so bombarded with new signals and craziness . . . A random person who walks around with a coin and flips it and then decides based on the outcome, that guy is going to have a horrible life."

Your life pattern is you. It's what you do, with whom, and where. It's the content that fills the vessel of your existence. A few decades ago this content was private, but also forgettable, a stream of experience that flowed into oblivion. It's now less private and the stream flows to someone's server.

Privacy hasn't diminished in *importance* simply because we're adding connections, embedding new sensing capabilities into our physical world, and using mobile technology in new ways. But our discussion of privacy seems to have remained in a static state of fretting for decades. One of the best defenses against potential misuse is to personally get hold of your data. This sounds like a chore because it is. We should demand that this become much easier in the years ahead than it is today. Your data, and what it says about your future, belongs to you first and foremost.

If you've got a bit of spare time, sign up to receive updates from the Electronic Frontier Foundation, subscribe to the ACLU's *Free Future* blog RSS feed, and when these organizations plead at the end of the year for funds, maybe send them a check. Just know that even the ACLU can't change the simple fact that privacy isn't what it was, and it's not what we imagine it to be. If we choose to give up the ghost, reconcile ourselves to the reality of modern interconnectedness, of greater visibility, of transparency, and thus of predictability, the question becomes: What's the upside to the fact that we create data in everything we do, and that makes us predictable? What can we predict about ourselves? What's the right way to live in the open?

CHAPTER 2
The Signal from Within

THE year is 2009. The setting is a television station in Washington, D.C. I'm about to do an interview with CBS correspondent Tracy Smith on how technology is going to change life in the coming decades. It should be a simple Q&A, but this is television and I've convinced myself I can't go to this interview without a prop, a physical relic from the future. I can pick maybe one object to represent the most important trend that will change life on this planet and whatever I pick has to fit in my pocket. Two days before my interview I start to panic. What the hell kind of world-changing, futuristic invention fits in your pocket?

I remember a story I had written for the *Futurist* the previous month about a California start-up that was manufacturing an interesting new health product, not a pill or some exercise contraption but a 2-GHz RFID clothes clip that could record calorie burn, sleep, and other biophysical signals. Such devices weren't exactly uncommon in 2009 but unlike a standard pedometer, this device let the user upload stats directly to a Web site for automatic tracking and potential sharing with doctors, family, friends, or even the public.

The gadget spoke to some of the novel ways people would soon be interfacing with computers outside the act of typing. The idea of consumers buying a device to track and publicly broadcast the sort of signals that would normally only be collected in a hospital setting suggested the emergence of an entirely new product class. *Normal* was shifting and this device was the evidence. It had the makings of a great prop for a future-themed TV bit.

In the segment that eventually aired, I'm about thirty pounds overweight with an Irish double chin. I hold the device up to the camera and reveal how the gadget will allow users to monitor and broadcast their activity levels. A voice-over reads, "Our descendants might wear their heart rates on their sleeves with a device like this, the Fitbit."

I know, I know. *Descendants?* Were we really that timid in 2009?

Since its launch, Fitbit has received considerable media coverage in the *New York Times*, *Time* magazine, and others. Not all of the publicity has been positive. In 2011, a tech blogger pointed out that many users who were confused or ignorant of the device's privacy settings were inadvertently sharing the data details of their sexual activity, prompting *Forbes* writer Kashmir Hill to observe that sex, "even at its most 'difficult' . . . doesn't burn nearly as many calories as the elliptical at the gym."[1]

The market for devices such as the Fitbit and its close cousins such as RunKeeper and various smart scales, armbands, and clips has expanded tremendously in the last decade. According to a recent report from the Pew Internet and American Life survey, one in five Americans use some sort of device to track health stats for either themselves or someone else.[2] These gadgets have become recognizable symbols of one of the most interesting health movements since cutting out carbs: Quantified Self (QS).

QS is either "sort of mainstream, now" according to movement cofounder and *Wired* maverick Kevin Kelly[3] or an unfortunate obsession afflicting certain young people "who like to record and share every aspect of their lives no matter how inconsequential," according to *Vanity Fair*'s Graydon Carter.[4] At the *Quantified Self* blog, the community's Plymouth Rock, *Wired* editor and movement leader Gary Wolf describes QS as "a collaboration of users and tool makers who share an interest in self-knowledge through self-tracking. We exchange information about our personal projects, the tools we use, tips we've gleaned, [and] lessons we've learned." While a lot of the data that self-trackers collect are directly related to health, much of it isn't. Any measurable behavior or experience can be part of a self-tracking regimen.

One young woman I met who embodies this movement is Sacha Chua. Here's what self-knowledge through quantification looks like: when Chua visits a clothing Web site she is impervious to impulse purchases. She can predict with statistical certainty which outfits and garments she is more likely to wear and which she is not. She knows how long she sleeps during a night, her average resting heart rate, and whom she is most likely to talk to on a given day (her mother is number two—a fact her mother hates). She has a remarkably clear idea of how long she is likely to spend on any given task. When she chooses to, she can determine—down to the minute—how wind speed and direction will slow her bike ride to work.

By day, she's a freelance technical consultant and boasts an impressive résumé with a master's degree in human-computer interaction and loads of Fortune 500 clients. By night, she's one of the cofounders of Quantified Self Toronto and editor of *Living an Awesome Life*, a blog in which she documents her experiments and gives tips and advice to interested novices.

We meet at a bar with several other members of her group. Though we are at an establishment that serves liquor she orders only water, which I find almost obscene. "I don't drink anything that might affect my brain," she explains. She also doesn't do caffeine. "I'm too excited already. They would have to scrape me off the ceiling."

When the subject hits upon her self-quant projects, she becomes talkative and bubbly, like a child at a party. Unlike Kelly, she doesn't feel socially encouraged to track the ins and outs of her life on this planet. She does it expertly, but knows this is not something that will find sympathy or interest in Toronto, Ontario. "Some people think it's a bit obsessive-compulsive, this record keeping," she admits. Chua doesn't match the profile of someone hobbled by a neurotic condition like obsessive-compulsive disorder. She isn't a dysfunctional person. Quite the opposite. She's *hyper*functional. She's happy and confident and seems more *awake* than regular folk—in the Buddhist sense of the term; not wired, just present.

I ask her if she has a particular goal that's she's striving to achieve through self-tracking. In fact, she says, she has thousands of small goals, each built around optimization. She wants to get better at commuting, better at making time for her husband and her child; she simply wants to use data and the technologies that make data generation easy to get closer to a scientifically perfect existence.

"It's the pursuit of relentless improvement," she says. "It's compounding interest in life."

The practice of collecting data about groups for public purposes has been around since the Babylonians began taking census surveys in 4000 BC. But collecting data about oneself hasn't caught on. There's a very simple reason for this: cost versus benefit. When a nation, state, or kingdom conducts a survey on national health or the activities of its citizens, the cost is shared across the public. When a corporation decides it wants to collect every piece of information that may ever be recorded by any machine anywhere, they, too, have the option of first estimating the potential value of the data they're seeking, then the price of obtaining it, and then the monetary value of publicizing the results of its exploration or keeping the findings private. But the benefits of systematically collecting *personal* data are more hidden. What are the rules for tabulating this stuff? How much should be stored? Most important, will the benefit of this self-knowledge ever surpass the cost in terms of time and energy spent collecting it? The answer is very subjective, as everyone understands costs and benefits in very unique ways—much like hope, fear, resentment, and envy.

Benjamin Franklin could be considered the American pioneer of self-quantification and his experiences provide a clue as to the *objective* value of personal data. In his autobiography he discusses how he became a fan of *The Golden Verses of Pythagorus*. These verses are little more than simple and sound admonishments to speak and act with reflection, avoid envy, bad eating, self-abuse, and so on. Though each verse is but a single line, taken altogether the seventy-one verses provide a sort of laundry-list guide on how to approach life. It's useful, though the modern reader will likely find there are far too many references to Jupiter. They inspired Franklin to keep a book of what he called thirteen "daily virtues." Franklin's virtues, like Pythagorus's verses, were extremely short, a list of commonsense watchwords and concepts to arm the individual against such eighteenth-century scourges of character as trifling and dullness. Number twelve, chastity, is accompanied by a proscription against the use of "venery [sex] but for health or offspring, never to dulness." Franklin's virtues were ordered sequentially with the most important first, beginning with temperance, which proscribed against doing anything to excess. It was this virtue that made possible all other virtues in Franklin's mind.

The book itself was nothing more than pages of tables, seven columns down (for each day of the week) and thirteen rows across for each virtue. At the end of the day, Franklin would look over the table and if he had run afoul of one of the virtues, he would mark the corresponding cell with a black dot.[5] In his autobiography Franklin includes a sample template but doesn't go into much detail about how well he performed, admitting only that when he began the experiment he was "surprised to find [him]self so much fuller of faults than [he] had imagined." (Reputation suggests that he ran afoul of the virtue of chastity rather often.) That Franklin was shocked to discover himself less virtuous than he believed is surprising when you consider how accomplished he was in virtually every area of life. How is it that a man of Ben Franklin's genius could be *unaware* of his own limitations?

In fact, this self-ignorance fits with what numerous psychologists have discovered about humanity's innate tendency to overestimate our own competence, fairness, and virtue. This tendency has been observed so many times it goes by several names, such as the Dunning-Kruger effect, after two psychologists who observed experimentally a natural human tendency to inflate our own level of competence in a variety of domains, both intellectual and social.[6] Then there is the more clinical term "anosognosia," or lack of self-awareness (reserved for extreme cases). Probably the best synonym for this predilection of self-ignorance—shared by all humanity but very rarely acknowledged—comes from Nobel laureate Daniel Kahneman, who throughout his work refers to it as the "inside view."

The inside view is the human tendency to predict success in novel endeavors—and the timing of success—as derived from a statistically insignificant reference class, namely one's personal experience. That may seem somewhat unrelated to Ben Franklin's problem of being virtuous until you consider temperance, chastity, and moderation not as innate qualities a person either possesses or lacks but as objectives to strive for daily, which is exactly how Franklin began to perceive them. The surprise he describes in his autobiography is not the fact that he is without virtue but that he thinks he falls well short of a level of virtue he believed he had obtained. He comes face-to-face with his own cognitive bias, his own inside view!

In describing what the inside view looks like, Kahneman will often rely on a personal anecdote from the 1970s. He was able to talk the Israeli Ministry of Education into creating an entire college-level curriculum around his area of expertise: judgment and decision making. He assembled a team with experience in teaching, editing, writing chapters in textbooks, and so on. One member had even designed curricula in the past. Kahneman asked these folks to estimate how long they would need to complete this chore of creating a curriculum and expected they would be able to answer with an above-average degree of accuracy. Keep in mind that these were highly rational human beings, top performers in their fields. The team projected they would have the task complete in two years, tops. Thing is, they knew statistically that 40 percent of similar teams had failed at similar tasks, and those that succeeded finished behind schedule. They couldn't convince themselves that the statistics applied to them. Kahneman's team took six years and the curriculum was never adopted.

So the inside view is human nature, but you may say the same thing about overeating, oversleeping, and chronic masturbation. It should be possible, then, to avoid it. This is where self-tracking comes in. Michael J. Mauboussin, chief investment strategist for Legg Mason, has written about Kahneman's work and suggests that this is a simple matter of summoning the will to find the *right* reference class. In short: "Assess the distribution of outcomes. Once you have a reference class, take a close look at the rate of success and failure. Study the distribution, including the average outcome, the most common outcome, and check for extreme successes or failures."[7]

Without exactly understanding the concept of reference class, Franklin knew that the first step to discovering his own capacity for virtuous behavior was first to objectively record how often he fell short of his goal. He needed better and more regular data. He kept up the practice of virtue logging his entire life and seemed to derive great therapeutic benefit from it. But aside from a certain regularity of record keeping, his conclusions about which virtues he had flouted were entirely subjective and imprecise.

Self-tracking became much more of an actual science with the advent of electronic computation; the ability to input, store, and process lots of data. Though it sounds faddish, QS has been around nearly as long as the modern PC. The home computer allowed for better record keeping of more activity at much less cost of time and energy and also provided a much better analysis of those records. This is why the two most significant celebrity devotees of QS are also two of the world's more important modern figures in computer science.

The Rigorously Examined Lives of Ray Kurzweil and Stephen Wolfram

Here's what most people know about Ray Kurzweil: he's a futurist and bestselling author who mainstreamed the concept of Singularity—a future tipping point in which technological progress accelerates at an exponential rate, allowing humankind to devise a means to cure all death and illness and become virtually immortal. Some people also know that he's a celebrated inventor, having created the world's first charge-coupled flatbed scanner, a brand of electric keyboards that bears his name, a device that can read text in books aloud for the blind, and a number of other programs and gadgets.

Here's what very few people know about one of the most important minds in AI: though he's sometimes accused of being a robot himself, he's actually very funny. It's a bone-dry, snap-quick, intensely human humor that doesn't at all resemble an AI entity. Few people also know that his father, a composer, died of heart disease at a relatively young age and passed his genetic predisposition for elevated cholesterol down to his son.

Kurzweil was also diagnosed with type 2 diabetes in the mid-1980s. The insulin treatment his doctor put him on made him a good deal fatter but no healthier. It seemed his congenital heart condition and treatment were locked in a race to kill him. So he decided to treat himself. Doing so meant constantly researching new ways to improve his health and carefully logging dozens of different biological reactions to drugs, diet, sleep, and exercise.

In 1999 he met his current physician and sometime collaborator Terry Grossman, on whom he made an immediate impression, as Grossman recounted for the *Futurist* magazine.

"I am not surprised when patients come to see me with a notebook of spreadsheets detailing various data extracted from their daily lives: blood pressure, weight, cholesterol, blood sugar levels, amount of exercise, etc., carefully tabulated for several years. But all previous data collections I had seen, even those organized into Excel and meticulously graphed, paled in comparison to Ray's. His data collection was so thorough and meticulous that he could tell me what he ate for lunch on June 23, 1989 (as well as what he ate for lunch every other day for several years before that time). And not only what he ate, but the number of grams of each serving and calories consumed, as well as the number of calories he burned that day through exercise—every day for decades!"[8]

Ray Kurzweil has been experimenting on himself and carefully recording the results for almost thirty years now. When I ran into him at a San Francisco event in October 2012, he told me that he credits self-quantification with helping him overcome the threat of heart disease and diabetes. At the very least, it's helped him outlive his father, who died at age fifty-eight.

"These numbers, you can change them . . . very radically. You're not a prisoner to this characteristic. I've dramatically changed who I am."

Improving health is, today, the most common use of self-quantification. But there are entire areas of life where a rigorous approach to data collection and analysis can lead to better outcomes; this includes analyzing communication, work, and buying patterns. Computer scientist Stephen Wolfram shares a lot in common with Ray Kurzweil. He too was an early pioneer of the use of personal computers to create and store records of virtually any signal, transaction, or change that could be recorded, even though he didn't know exactly what he intended to do with that information when he started collecting it.

Wolfram is best known as the creator of the Wolfram Alpha search engine. It's similar to Google except that the inputs—what you type in the search box—and the outputs are mathematical, sort of like a calculator that can read the Web. Unlike the Google engine, which responds to every question with a ranked list of pages, Alpha actually computes specific answers.

When you use Wolfram Alpha, a lot of queries come back with results like "need more information," but when it works, it's miraculous. Need to know the average life span for a human being in France? Alpha can give you the number in years (81.4) and a breakdown for the percentage of the population dying older or younger, as well as a graph for how the number has changed over time. For instance, there's a dip during the two world wars, but a strange blip upward in the middle of both. Need to predict how long it's going to take you to read 1,000 pages in a standard textbook? The answer, via Wolfram Alpha, is thirty hours, assuming this 1,000-page textbook is perfectly statistically average, contains 3.05 megabytes of information, 500,000 words, 45,000 lines, and 5.1 characters per word.

In March 2012 Wolfram surprised the world with a revelation on his blog that he had been Wolfram Alpha–izing his own life for close to thirty years.[9] He had a log of not only every phone call he had made in that time but *every keystroke* since 2002 (more than 100 million), and every e-mail he had sent since 1989 (more than 200,000). He knows, roughly, how many steps he's taken in the last two years, according to GPS and room-by-room motion sensor data. He has all of his medical test results and his personal genome, which he describes as "not yet very useful." If it's a Thursday night and it's 10 P.M., he knows there's a 50 percent probability that he'll be on the phone. When I had the opportunity to speak with him I wanted to know what he's learned.

"One thing I've found out is that I'm far more habitual than I ever imagined," he told me. For instance, he now knows that if he ignores his in-box for five days, it will take him two weeks to get up to date. He has a formula to estimate the likelihood that he will procrastinate on a given project, the average time of procrastination, and the best way to avoid it.

"I'm one of these people that for certain types of tasks, there's no point in starting early; I won't finish it until just in time anyway. I have to know how long it's actually going to take to finish so that I can get it done in an efficient way. If I start it too early, the task expands into the available space," he says.

Wolfram hopes that one day he'll be able to use this painfully extracted self-knowledge to figure out when he's at his creative peak. This is not an insignificant matter, as his life is devoted to solving problems that are supposed to be impossible. "What I've had to do is figure out an efficient scheme to create on demand. I've gotten better at it. I know that if I think about something for a certain period of time, a window will appear, a period of a few hours when I'll either have a good idea or I won't."

Wolfram's experience speaks to one of the more important, near-term applications of QS. In the decade ahead, a lot more people will be tracking themselves to guard against something like the inside view.

But Wolfram and Kurzweil have created what Ben Franklin couldn't: a reference class of personal data that meets the criteria of "large enough to be significant." And they did so at a cost of time and effort that—while still high—was lower for them than it would be for someone else. As a result, they are perhaps better armed against the inside view than the rest of us.

If you're not exactly eager to tabulate every step you take and every gram of sodium you ingest, some recent research suggests that you can greatly improve your health by simply watching just one signal: how you react to stressful events.

In 2012 University of Pennsylvania researcher David Almeida and some colleagues published a paper showing that the most important predictor of a future chronic health condition (aside from smoking, drinking, and engaging in conspicuously unhealthy behavior) was overreacting to routine, psychologically taxing incidents. When they interviewed subjects about how little stressors such as car breakdowns, angry e-mails, small disappointments, the little annoyances of modern life, affected them emotionally, they found that "for every one unit increase in affective reactivity [people reporting a big emotional change resulting from the stressful event], there was a 10% increase in the risk of reporting a chronic health condition 10 years later."[10]

The researchers didn't find that people who were *exposed* to more stressful experiences were more likely to develop a chronic health condition. Rather, the increase was isolated to people who reported feeling very different emotionally on a day that they encountered a stressor than on a day when they did not.[11]

This is a classic inside-view problem. Very few people keep track of how they react to little stressors. The costs of keeping such a record, in terms of inconvenience, are too high. Yet hidden in those reactions may be powerful clues to our future health. If it were easy and cheap to keep that data around, and if we were able to make sense of it quickly, we would surely keep a log of how stressed we felt at any given moment.

When I asked Kahneman via Skype at the Singularity Summit 2012 what he thought of the self-quant trend, he was guardedly optimistic about the potential applications of quantification techniques for physicians. Adopting the outside view will never be intuitive, he said. "But at least in principle there is an opportunity for people to discover regularities in their own lives. There will be an opportunity to look at the outcome of similar cases . . . A physician could have intuitions about a patient, but supplementing that *intuition with instantly available statistics* will likely result in fewer mistakes" (emphasis added).

Though Stephen Wolfram is the seminal self-quantifier, he believes that personal data collection has to become both a great deal easier and more immediately rewarding before it becomes mainstream. The technology itself is no longer the problem. What's lacking is expertise. We imagine computerized data to be neat and clean. But getting our data into a usable form isn't as simple as just pressing the Return key.

"There's all sorts of plumbing that has to be done," says Wolfram, meaning that service providers need to make these services easier to use for everyone, not just computer geniuses. "If everyone used one vendor's equipment, that would be a start. But that won't happen. It will stay a complex, multiproduct, multivendor environment. This data [is] in computers; [it's] in pedometers; [it's] in lots of different places. First step is to make it possible to upload to somewhere. Maybe upload it to a cloud that's shared but some people are too paranoid to do that. I don't think there's anything *technically* difficult about this. It just requires all sorts of *work*."

In the past couple of years, some enterprising start-ups have sprung up to relieve the amount of work involved in keeping track of signals, physical states, and so on. The United Kingdom's Tictrac is a platform that allows you to take your data from different devices and sites and create a snapshot of yourself in the present. I spoke with founder Martin Blinder while the company was still beta testing. It's since opened to the public and has been steadily gaining users. He knows that Tictrac will only succeed if it can offer personal analytics in a way that's intuitive and user-friendly. It needs to be able to take your data and present you with a future prediction you couldn't have reached yourself, and do it in a way that's perfectly understandable at any given moment.

"We want to offer a breakdown of the food you've eaten in the last month, type of food and calories, and whatnot. But we also want you to be able to see analytics across different data sets, so you can pull in your calendar so you can see that you spend about twelve hours a week on business lunches—and then cross that with weight and find a correlation between the two," he told me.

The average Tictrac user has as many as ten thousand different points of data that users bring to the practice of lifestyle management. Those points include everything from Facebook posts to e-mails to GPS or fitness logs of physical activity from such devices as Fitbit, Nike+, and Blipcare. Some of his more exceptional users have twenty thousand data points. And the exceptional is quickly becoming the average. As a measure for how much personal data there is today versus ten years ago, only two digital data streams that Tictrac users build into the graphs were around prior to 2003: e-mail and calendar.

Perhaps the most encouraging aspect of the Tictrac platform is that the number of data sources and possible insights is limited only by how much data can easily be collected through an API. With enough streaming data, it's possible to see how all of your life areas interact, how overbooking appointments affects your exercise levels, how your communications with one person changes your drinking and sleeping or monthly expenses, even how what you eat influences your electricity usage. These sorts of data streams will grow as we integrate more sensing and broadcasting capability into more objects and our environment.

Devices like the Q Sensor from Affectiva, which looks like a strange wristwatch, can measure your level of interest and engagement in a given activity based on electrodermal conductance (i.e., how much electricity your skin is emitting), and in 2012 a company called Cogito, founded by two MIT grads working under a Defense Advanced Research Projects Agency (DARPA) grant, created a platform that can detect—and predict—your mood levels based on tone and cadence of speech. It was intended to help mental health professionals working with returning vets better anticipate the rise of depression. One day, a future version of Cogito could make its way into an iPhone app capable of helping you anticipate and plan around your future feelings. It would telegraph your *emotional* future the way horoscopes are supposed to but would be based on data points accumulated through actual life, as opposed to unproven notions about the effects of planets. There are countless other examples of self-diagnostic apps waiting to be developed. Research from University of Maryland scholar Lisa M. Viser has shown that it's possible to detect dementia in keystroke patterns, or simply based on changes in the way someone types over a period of weeks or months. (You can also detect whether someone's been drinking at lunch.)[12] Bream Brush is a smart toothbrush that dialogues with the user's smartphone via an app to keep a log of brushing time. The data can be shared with the user's dentist, insurance provider, et cetera.[13]

Let's assume one of these start-ups, or one not yet conceived, makes it to mainstream adoption. Once that occurs, the personal costs for self-quantification will have collapsed in just a few decades. In the 1980s, when Ray Kurzweil decided to flout the advice of his doctor, take himself off insulin, and begin keeping a detailed log of every meal he ate and what was in it, few other people would have had the patience, know-how, or inclination to attempt anything similar. The behavior at the time seemed positively bizarre. When Kurzweil first began his self-quantification experiments, the costs in terms of time and effort were a bit lower for him than they would be for anybody else, except Stephen Wolfram. Today, they're joined by enterprising people such as Sacha Chua, and the numbers are growing.

We are one app away from becoming Ray Kurzweil.

Here's what that app might look like to you in practice. You would give the program access to your biophysical signals, gleaned from your activity levels, mood analyzers, implants if you have any, e-mail and voice mail, et cetera. The program in turn would give you a rapidly evolving window into your future health. On any given day, you might receive a notification with the following warning: "Dear Patrick, as a result of that stress event you had a couple of weeks ago, the dizzy spell you complained of last night, and the fact that you've recently increased your daily alcohol consumption from two glasses of Merlot to four, your probability for stroke in the next year has just increased to 10 percent."

Naturally, if you received this message, you would act to avert this stroke before it happened, rendering the prediction incorrect, but still invaluable.

Yet more cloud processing and an abundance of carefully collected personal data aren't the magic ingredients that are going to bring the above scenario to life. Even with the right technology and a seamless interface or analytics engine to take the difficult work out of making usable predictions from your data, the most important component of your changing health picture is other people's health data. Here's the trade-off, the point where our outmoded ideas of privacy begin to get in the way of progress and better health.

The Network Is Your Doctor

In June 2012 a group of researchers from MIT and Columbia created a system that can predict future illness. They call the system the Hierarchical Association Rule Model or HARM (a bizarre but at least memorable acronym for a medical algorithm). It can't tell you what's wrong with you right now; instead, the algorithm determines what you're likely to get *next* on the basis of a current diagnosis combined with certain demographic features such as race, age, and so on. To build it, they used clinical trial data from around 42,000 patient encounters and 2,300 patients, all at least forty years old.

These were people who had signed up to test new medicines so their medical records were more thoroughly filled out than is the norm. The subjects were also encouraged to report back on what they were feeling and experiencing, as such data could have an effect on the drug's marketability. What were the results? It turns out stroke is more predictable than many statisticians had believed, even though the proximate causes for stroke are still difficult to determine.[14]

The system works by finding correlations among thousands of patients sharing their history, not by looking for causes. This is a big departure from the way medicine is traditionally practiced and taught. It also only works when thousands of people elect to share their most personal data.

Future breakthroughs in the application of highly personalized data to medicine will depend tremendously on a willingness to share that sort of information. The question is how to do it in a way that doesn't come back to haunt you. From a researcher's perspective, the problem becomes one of how to build privacy controls that allow your users to share but that allow your model, algorithm, system, or Web site to access the most valuable information.

The fledgling Consent to Research project headed by John Wilbanks is a great example of an organization that fully understands the importance of sharing health data for future medical practice but also understands the risks people take in exposing themselves. Wilbanks points out that more than one in ten people in the United States have a rare disease (defined as a disease that fewer than two hundred thousand people are diagnosed with per year), have a family member with a rare disease, or have a first-degree friend with a rare disease. There are a lot of illnesses that very few people have, but the sheer number of them affects us all.

Wilbanks's sister is one such person. "Our best guess is that she has some kind of psoriatic arthritis. We don't know what kind," he told me and a few other folks at a Singularity Summit event in San Francisco. "The insurance industry is already quite good at denying her care based on the actuarial tables. So we pay for PET scans out of our own pocket."

This, in part, is why Wilbanks views the idea of sharing medical data a bit differently than other people do. As far as he's concerned, a system that keeps your information safe but can't find a cure for your sister's disease doesn't work for anyone. "In our family, the potential risk of sharing our data is low compared to the potential benefit that might come."

One of the fundamental flaws of the big data present, as opposed to the naked future, is that the value or benefits of sharing data is experienced collectively but the risk is experienced *personally.* Wilbanks is trying to get people to share their medical information in a way that creates "data bait" for researchers to solve problems that afflict families like his. "That's a better investment strategy as a nonwealthy family than paying for research into psoriatic arthritis," he says. "There are people who are good at math who would like to work on something besides hedge funds, but the transaction costs for getting at health data are so high that they won't bother."

This is one of the more egregious dysfunctions of our healthcare system: getting good and useful medical information on a statistically significant population of subjects has never been more expensive. Researchers have estimated the cost for getting patient consent to review medical records at $248 per person.[15] That's part of the reason why the cost of a clinical drug trial for a single medication can run as high as $5 billion, according to Tuft's researcher Kenneth A. Getz. This is an insane amount of money pointing to an artificial scarcity of health data because the *availability* of potentially useful health data has never been greater. It's not just locked away in blood tests but, again, in our phones, our actions, the daily map of comings and goings.

However, Wilbanks agrees with Wolfram that more data won't equal better health by itself. "The more we measure ourselves, the more stressed we'll be. If we don't figure out ways to get that data into the sorts of models that tell you how likely you are to respond to a drug, we'll just be creating more noise."

We would be well served by a health-data movement that's open and user powered, much like the Internet itself. This is what he hopes to offer with the Consent to Research project. "I'm arguing for something open at the core, closed at the edges, and innovative without centralized control."

Open, innovative, and decentralized are three adjectives that don't apply to health care today. But this is no one's fault but our own. In the twentieth century people consumed health care in the form of hospital visits and medication. The occasion for this consumption was sickness or the symptoms of sickness. In the twenty-first century we have added greater emphasis to the prevention of illness. Diet, exercise, yoga, and wellness have become recognized keys to a long, healthy life. This reprioritization has been a positive step but not wholly transformative. We have remained a customer base. Health care is something you buy from a professional and health is something you make with supplies from Whole Foods.

The next phase of health care—should we choose it—will demand that we become active participants not only in our own health but in the health of others. The data we create has value beyond how we use it to jog more, diet better, and maintain height-weight proportionality. It's more valuable than the results we realize in the singular. My health reveals something about your health, and how you experience health is relevant to me. The next stage in the evolution of health care demands a blurring of the lines between doctor, patient, and researcher. When information about our habits as individuals can paint a full picture of how we are living and dying as a species, we all thrive better individually and collectively. That becomes more apparent when we consider how sharing information will become the best defense against sharing a cold.

CHAPTER 3
#sick

THE year is 1374. The Black Death is ravaging Europe. At the port of Venice, a ship is approaching from the east. It is met by a small *cammelli* boat, a type of flat-bottomed vessel designed to guide larger ships through the shallow waters of the Venetian lagoon. But this *cammelli* carries a special passenger, one of Venice's three newly appointed Guardians of Health. He boards the tall vessel and quickly sets to work inspecting the crew and cargo where he finds a man who is delirious with fever. The lymph nodes beneath the man's jaw are red and inflamed. The guardian informs the crew that they must leave port. Venice has just instituted an unprecedented protocol: any ship suspected of harboring men infected with the plague must move on for a period of *quaranta giorni*—forty days.

This edict marks the birth of the term "quarantine."

Why forty days and not fifty, or sixty, or a year? The specificity of this prescribed interval dates back, again, to the ancient Greek thinker Pythagoras and his Doctrine of Critical Days, which suggested that illness in the human body strengthened or weakened on the basis of natural cycles and rhythms. The most critical of these critical days was day forty, after which point if a diseased person had survived the terrors of tremors of her affliction, she was deemed safe for reentry into society. Many contend that Pythagoras believed the cycles of the moon were a key determinant. As the moon waxed and waned, so, in theory, could sickness.

Pythagoras's doctrine, which went on to influence the ancient surgeon Hippocrates who first turned it into medical practice, was probably an improvement over no system at all but the degree of that improvement is hard to determine. Despite the *quaranta giorni* policy, the Black Death would go on to ravage the city of Venice in the years ahead.

Skip to the year 2020. Twelve-year-old Josh Grant is in the school nurse's office. He doesn't know why.

"I feel fine," he tells the nurse (let's call her Nurse Gwen).

"That may be so, but you spent a good hour sitting six yards away from your girlfriend, Jessica Stickler, this morning. There's an eighty percent probability that she's going to come down with flu symptoms in the next day. Nothing to worry about. We'll genotype it."

"She's *not* my girlfriend," Josh answers. In truth, he's not yet sure if Jessica Stickler is his girlfriend. The prospect fills him with dread. "Anyway, that doesn't mean I have the flu."

"I know." Nurse Gwen tries to sound consoling. "Based on the exposure time, there's really only a ten percent chance you're a carrier. But if you are, then there's a sixty percent chance you'll infect Tim Miller during chemistry class later. And Tim's mother has nonrefundable tickets to *Disney Atlantis* this weekend."

"What does Tim's mother have to do with it?" asks Josh.

"She called us and asked us to pull you out of history class. Normally we would never pull a ten percenter out of class to scan for flu, not when I have ten forty percenters from second period alone." She laughs and then turns suddenly serious, like a faucet moving from open to shut. "Tim's mother can be rather assertive . . ."

Josh pulls out his phone and learns from Wikipedia that "the genotype of an organism is the inherited instructions it carries within its genetic code."[1] Before he can consider how that applies to him (or if Wikipedia is even the best source), Nurse Gwen asks him to open his mouth. Soon she's swabbing the inside of his cheek. She scans it using her phone's camera. A moment later the sequencer app makes a *ding* noise, indicating it's done analyzing the influenza strain. Nurse Gwen opens the report on her phone. Her eyes widen and her jaw drops.

"Oh my," she says in a frightened whisper. She rushes to a cabinet and pulls a small vial from a box marked TAMIFLU #39 and scans a bar code on the side. She next takes a surgical mask and puts it over her mouth and nose.

"What's going on?" asks Josh.

"We're just taking an extra careful look at your flu," answers Nurse Gwen.

"I don't have the flu," Josh repeats. "I feel fine."

He's no longer so sure of this.

From the word "genotype," Josh assumes that there must be something in the genetic code of his flu that's causing Nurse Gwen to act this way.

His mother storms into the nurse's office a few minutes later. "I just got a push notification that Josh is now at one hundred percent probability for flu and he was at ten percent an hour ago. What kind of practice do you run here?"

Nurse Gwen hands Josh's mother a mask.

"Put this on," the nurse commands.

Josh is now terrified.

"But it's just the flu, right?" asks Josh.

"It is the flu," says Nurse Gwen, "but it's a genetically novel flu strain." She pulls up a 3-D map of the earth's surface crisscrossed with bright lines connecting points in Asia, Eastern Europe, and the United States. At each point, the color of the line changes.

"I presumed that Josh had the flu we've been tracking for weeks. It's a fairly common one related to H3N2VVI. It originated in the People's Republic of Georgia two years ago and has gone through several small mutations, here in Hong Kong on December fourth, and here in New York on January fifteenth. Each time a mutation event occurs, the trajectory line changes. The next mutation event was forecast for February nineteenth. Based on the traffic patterns, we believed that this mutation would occur in Mexico City. But Josh's flu isn't H3N2VVI; it's H3NBX. This flu *was* limited to exotic avian megafauna. With Josh, it appears to have made a transgenic leap from birds to mammals."

"What's an exotic avian megafauna?" Josh feels a distinct sinking feeling.

"Harpy eagles, in this case. Most likely from Central America. They're often smuggled into the States."

"My twelve-year-old son is not a harpy eagle smuggler." Josh's mother is adamant on this point. "Look at his Twitter trail."

"Well . . ." says Josh.

The adults turn to him. His mother becomes very pale.

"The other day, when I was at Ray Bremmer's house tweeting about *Zombie Warz* for Xbox, I wasn't actually playing Xbox. Ray's uncle just got back from Panama with these birds and Ray said I could see them if I promised not to tell *anyone* . . ."

The sequencer app makes a bell sound. Nurse Gwen checks her phone for the report. She removes her mask. "Good news, it looks like this strain is very mild and isn't Tamiflu-39 resistant, at least not yet."

Josh's mother seems not to hear. "Of course Ray Bremmer would give you some wild bird flu . . ."

"Ray didn't give Josh this flu," Nurse Gwen clarifies. "The bird did. The more important question is where the flu is headed now. Looking at your geo-tagged posts, it appears you hung out with David McGill yesterday after the final bell and then with John Brooker this morning at recess. David's posts are now reading that he's at the airport with his father on his way to California. Running the projection map, it appears the Josh Grant flu will show up in Los Angeles tomorrow and from there it will be in Europe, Asia, and Africa by Thursday."

"Wait," says Josh in horror. "It's called the *Josh Grant* flu?"

"Just for now," says Nurse Gwen. "You are the first human to get it and I have to call it *something* when I talk about it with the CDC. I remember not long ago when people contracted the flu they would say they had a bug that was *going around*. Today, we can actually see the bugs and how they spread on a person-to-person level. We can name specific influenza strains after specific mutation points. Every flu can have a name. You are the first person to get what will be a fairly common flu in a couple of months. People will get your flu on subways, on airplanes, on cruise ships, and, of course, in classrooms. The Josh Grant could well be *the* seasonal flu next year. But not to worry, a couple of days of bed rest and liquids and you'll be fine."

"But *why* name every flu? Why track it if it's not dangerous?" Josh asks.

"By tracking the Josh Grant flu we can keep it from *becoming* dangerous. We can predict how many vaccines we'll need and at which point Tamiflu 39 will cease to be an effective treatment, necessitating the development of a Tamiflu 40, which reminds me . . ."

Gwen leaves the conversation to pick up a few extra shares of a pharmaceutical company that produces influenza medication. Josh's mother calls her workplace to take the rest of the week off as Josh collects his book bag and shuffles toward the car. His body will recover from the Josh Grant flu in two days. His relationship with Jessica Stickler, sadly, may not survive.

PERHAPS you aren't yet convinced that the naked future offers any improvement to compensate for the sacrifice of privacy that it demands. Certainly, this new era will distribute rewards and punishments unfairly and unequally (sort of like the Old Testament depiction of God). But consider that every year millions of people in the United States truck themselves down to clinics for flu shots and wind up getting the flu anyway. According to epidemiologists, flu shots are 70 percent effective in the general population at most.

The reason? Every shot contains an (inactive) mixture of only the three virus strains that epidemiologists believe are going to be prevalent in the coming season.[2]

In the last several years, that has included strains of H3N2 (the base of the swine flu virus and several other influenza strains common in mammals), H1N1 (the famous bird flu), and a variety of influenza B strains, which are considered less dangerous and more likely to strike later in the flu season. But this is a small percentage of the types of flu known to be in existence.

The Centers for Disease Control and Prevention (CDC) almost apologetically states on its Web site, “It’s not possible to predict with certainty which flu viruses will predominate during a given season. Flu viruses are constantly changing (called ‘antigenic drift’)—they can change from one season to the next or they can even change within the course of one flu season. Experts must pick which viruses to include in the vaccine many months in advance in order for vaccine to be produced and delivered on time.” Perhaps it’s a sign of how far medicine has advanced that we, like naive children, simply assume the shots we get will actually work.[3]

In the last several years, the emergence of superlarge, publicly accessible databases of virus sequences such as the Global Initiative on Sharing All Influenza Data (GISAID)[4] and the National Institutes of Health’s GenBank[5] have greatly reduced bureaucratic barriers to finding and sharing the most current information about new influenza observations.

Wider use of sequencing technology could lead to earlier detection of new types of flu, which would help pharmaceutical companies create better vaccines. Today, devices like Life Science’s Ion Proton can sequence all 3 billion base pairs of the human genome in less than a day for a price of $1,000, according to the machine’s makers. With just eight ribonucleic acid (RNA) segments, influenza is an exponentially simpler organism to sequence than the human genome. But sequencing influenza is very rarely done at a nurse’s office—what epidemiologists call “the point of surveillance.” Instead, when flu samples are collected they’re usually sent to a county or state public health lab, by which point a great deal of time has been lost.[6]

Collecting samples from birds and animals that are showing flu symptoms is, arguably, a more important step in curbing the spread of new deadly flu types. But that sort of sampling doesn't happen very often. As the editors of *Nature* pointed out in a recent Op-Ed:

"Just 7 of the 39 countries with more than 100 million poultry in 2010 collected more than 1,000 avian flu samples between 2003 and 2011. Eight countries—Brazil, Morocco, the Philippines, Colombia, Ecuador, Algeria, Venezuela and the Dominican Republic—collected none at all . . ."[7,8]

The current state of flu detection leaves much to be desired. Yet the Josh Grant scenario outlined above could become reality within a decade. You can see its initial outlines today.

Supramap

The date is Tuesday, April 10, 2012. Bioinformatics professor Daniel Janies is at the NIH in Bethesda, Maryland, to discuss his creation, an interactive map that plots the spread of specific flu strains around the world. It also allows researchers to draw inferences as to where those strains will go next and how the strain will evolve. To do this, the Supramap, as it's named, makes use of the details that often find their way into reports epidemiologists write for public health officials but that don't always seem important to people who aren't epidemiologists. This includes such facts as when the variation was first spotted and whether the host organism was wild or farm raised. "We can put all that information together because all kinds of information [have] a time and place," says Janies. "Every point on the earth is an observation, every point in space is an inference, not just an evolutionary midpoint, as a phylogeneticist would create, but also a *geographic* midpoint" [emphasis mine].

The map on the display behind him shows the progression of the H1N1 strain out of Asia. A series of lines are shooting up and across the earth's surface; some are close together, others stretch over continents. Most are bluish green. They come together to create what looks like a strange scaffold on top of a Google Earth map, linking Indonesia to Korea and Japan. At various junctures, the green lines turn white, then red, and arch higher and higher. The lines represent the movement of chickens, ducks, geese, humans, et cetera, across Asia and tell the epidemiologists which sort of animal brought what strain of the flu where. A goose brought it to Guodong, China, in 1996; a chicken—transported via a person—brought it to Indonesia after that.

Janies explains that the color changes are junctions where the flu virus skipped to a new host, indicating a mutation event. The taller the scaffold beam, the greater the degree of uncertainty in inference. There are a total of 1,528 steps in this particular progression of flu.

How does better and faster field reporting of new flu strains change this map? It creates more certainty about where and when mutation events occur, bringing the lines that are up in the atmosphere closer to the earth's surface. That allows for better prediction about where the flu is going next. One of the key drivers is the growth of computing power and cloud services. Virus paths and transgenic events can be simulated just like weather patterns or human interactions. As more simulations are run, confidence improves.

For instance, since 1993 epidemiologists have suspected that the comparatively dangerous H5N1 influenza virus's polymerase basic 2 protein (P2B) contains lysine at position 627.[9] That factoid may not sound particularly important to your future health but it is the presence of lysine in that position that allows the virus to reproduce in mammal lungs, which, at 33 degrees Celsius, are colder than bird lungs at 41 degrees Celsius. When a virus can reproduce in lungs, it goes airborne. The host begins to spread it through coughing and sneezing. If you can figure out under what conditions that particular mutation arose, what the weather and air were like, you can determine when an already deadly flu strain becomes much more spreadable.

In January 2012 a team of scientists from the University of Wisconsin successfully created a strain of H5N1 influenza with lysine in the 627 position of the P2B.[10] They demonstrated that the influenza could be passed between ferrets through coughing and sneezing. The government supported the work through grants. But the researchers' findings were placed under a moratorium for months, as many believed the research could serve as a manual for weaponizing H5N1. Following the publication of the finding, conservative members of the U.S. House of Representatives threatened legislation to restrict publication of similar research in the future. This presented a somewhat ironic situation: the researchers were working, in part, under government grant, and the government was trying to suppress the very research that it had funded.[11]

Controversial and expensive lab experiments like the one described above will remain important in figuring out how flu moves from one animal to another. But the research isn't cheap or easy, is potentially vulnerable to political infighting, and is certainly not fast. Incredibly, for decades epidemiologists suspected the link between lysine, the P2B protein, and aerosol communicability of H5N1 in mammals, but it took until 2012 to show exactly how that mutation occurs.

The United Nations has forecast that an H5N1 pandemic could kill between 5 million and 150 million people around the world. As researchers Tyler J. Kokjohn and Kimbal E. Cooper point out, "The large discrepancy between those two figures reflects the tremendous difference that preparation could play in facing a global pandemic. Fatality rates will vary depending on the strength of the resources that the national health institutions have in place."[12]

The best guard against the worst possible scenario isn't just more data; it's field data broadcast in real time. Databases like ProMED-mail and Global Infectious Diseases and Epidemiology Network (GIDEON) have made some disease clusters much easier to remotely detect, in part because they enable the spread of information not just between local health-care workers and global health organizations but also between clinic workers in the same area who may be on the front lines of a potential outbreak and not know it. This emerging capability is particularly important in

Southeast Asia where diagnostics is often very difficult but where the climate and population conditions are ideal for epidemics.[13]

Genomic and RNA sequencing have become steadily more cheap and ubiquitous, advancing even faster than computer information technology. As Scripps bioscientist Eric Topol points out in his book *The Creative Destruction of Medicine*, in the 1970s sequencing was limited to ten thousand pairs at a cost of $10 per base. But in the 2000s, sequencing machines capable of reading "hundreds of thousands" of bits of genome code in parallel increased the speed and decreased the cost of sequencing "a thousandfold."[14]

Today, RNA sequencing of flu strains requires nucleic acid sequence-based amplification machines and polymerase chain-reaction devices that are usually about desktop-size and require a bit of technical training to operate.

Now imagine these two capabilities merging. Hospitals around the country and around the world already use handheld diagnostic equipment, sometimes called point-of-care tests, or POCTs, to *detect* tuberculosis, influenza, and a handful of other illnesses. In the coming years POCTs will be the leapfrog technology that brings twenty-first-century diagnostics to remote village settings, little one-road towns where no one ever thought it affordable to even build a clinic, much less a diagnostics lab. Not surprisingly, the Gates Foundation (with Qualcomm) today offers tens of millions of dollars in grant money to support the design of better POCT machines. There's no technological barrier that stands in the way of putting an influenza detector in the hands of poultry farmers. Turning these detectors into flu genome *sequencers*, transforming a big data stream into a sensed data stream, is also within our reach. Future breakthroughs in our understanding of microfluidics, better sensors (through nanotechnology), wider broadband coverage, and growth in cloud computing will enable anyone to take a flu sample, digitize the gleaned data immediately, upload it to the cloud for sequencing, and report flu mutations, all with the press of a button.

Imagine a naked future in which handheld devices and apps have brought not only diagnostic capability but advanced sequencing capability into the places where disease hot spots are most likely to flare up; where every poultry worker, school nurse, office or metro-station manager, mother, father, or student carried a handheld sequencer on their phone; a time when an army of people saw disease surveillance as part of their job, the same way we all feel a certain duty to report a suspicious package we see left on a train or in an airport.

"I see a future where commodity sequencing will be deployed widely in doctors' offices, in public health institutes, or maybe in the field in autonomous devices," Janies told the NIH assembly. "Every night I would like to calculate a new map and show it to whomever needs to see it."

But flu is more than just chemical reactions occurring inside the bodies of ducks, geese, pigs, and people. Behind every person-to-person flu transmission is a story of two humans connecting. Those stories are essential to better predicting flu movement.

Finding Your Flu Triangle

Schools are to the seasonal flu what gasoline is to fire, yet very little is known about how people actually move in these sorts of enclosed, highly populated spaces. For instance, let's say you wanted to slow the spread of influenza through a school. You could start by inoculating the most connected individuals, the people with the widest circle of associations. These would be teachers (connected by virtue of their official role) and popular kids (who earned their connectedness through blood conquest). If you could isolate these people at the start of an outbreak, you could halt the spread of flu in that population. Right?

Marcel Salathe of Pennsylvania State University set out to test this notion. He outfitted 788 U.S high school students with small sensors that recorded their movements at twenty-second intervals. Salathe and his team recorded 762,868 close personal interactions (an interaction at a distance of about ten feet, which is about the distance at which people mouth spray one another with flu). After he collected a day's worth of data he and his team ran one thousand simulations for each participant. The results: about 70 percent of the time, the infected will isolate themselves from the rest of the population and go home, resulting in no major outbreak. But the remaining 30 percent of simulations showed the contagion spreading. In some scenarios 1.3 percent of the school population went on to be infected; in others (for H1N1) the number was closer to 50 percent.[15]

The popular kids and teachers didn't spread the virus any faster than anyone else. The average close personal interaction in an American high school, according to Salathe's research, is a bit less than five minutes. But it's chopped up into an average of eighteen twenty-second intervals of hallway shoulder rubbing, lab partnering, cafeteria chatting. What was most surprising was that everyone rubs shoulders with everyone else. Separating or inoculating one group of *seemingly* more connected people had no major effect. So you aren't more likely to get the flu from the popular girl simply because she has more hangers-on, thus more potential infectors. You simply don't spend enough time with those folks. Salathe's model shows that the chance of getting the flu from a random person in your high school is 1.35 percent because your exposure to random people is just about five minutes per person. You're most likely to get sick from one of your two best friends, people with whom you share a lot of breathing and touching space.

Nicholas Christakis and James Fowler describe in their book, *Connected: The Surprising Power of Our Social Networks and How They Shape Our Lives*, for a typical American there's a 52 percent chance that any two people in your social network know each other. Christakis and Fowler call this transitivity. Flu is often passed from contact with various surfaces but when it's passed from person to person, transitivity is the primary factor, specifically triangular transitivity. Kids in high school spend a lot of time in groups of three, or closed triangles.[16]

If you know who is in your triangle then you know one of the two people most likely to get you sick. You can calculate the probability of that person infecting you on the basis of the amount of time you spend with that person, or colocate with them. The odds of flu transmission between people in close contact rises by a factor of 0.003 every twenty seconds, something a group of Johns Hopkins researchers discovered in the 1970s after watching how flu spread on airplanes. If you know who is in your triangle, and you know the amount of time you're going to spend with them, then you can put a number on the odds of getting their bug (assuming you don't do anything foolish like touch something they've first touched). This score would be far from perfect. But it need not be perfect to change in behavior and potentially limit the spread of influenza.[17]

But having the formula isn't enough; you want to be able to compute the probability of getting the flu from someone before you go out and colocate with them. This is where social networks are creating a naked future for communicable disease.

The Figurative Flu

For computer scientist Michael Paul, Twitter is much more than a social networking site. It's a window into the physical states of millions of users around the world, and potentially a boon to future public health. Between May 2009 and October 2010, he and his research supervisor, Mark Dredze, created a program to classify "sick" tweets in terms of various symptoms being expressed. They called it Ailment Topic Aspect Model, or ATAM.[18]

Every week, the CDC releases new flu numbers. The researchers discovered their ATAM model predicted official flu numbers before the CDC released them. They were able to work with the tweets daily and could have gone faster. The CDC method for collecting, analyzing, and releasing data is comparatively glacial. As Paul explained to me at his lab at Johns Hopkins, "The CDC takes two weeks because all of the CDC's tools are limited. They survey hospitals. They literally call around and ask for numbers on positive specimens that week. Getting that data takes a while. The numbers for the flu rate in a week come out about two weeks after that week. If you want numbers for an outbreak of a novel epidemic, like SARS, two weeks is too long."

Paul and Dredze aren't the first research team to attempt to use data on what people are typing into their computers and phones to forecast illness rates. Google Flu Trends famously can also predict influenza rates before official CDC reporting. But this flu trends program is query based. Search-engine queries reveal only what people are interested in, and the origin of that interest is a matter of guesswork, easier in some cases than in others.

If a lot of people in a particular area are querying "flu shots" that's a *fairly* good indicator of a flu outbreak, but it's also an indirect indicator. By definition, a query is a question, not a statement. We have a good but not perfect sense of why people are asking about flu shots. What if you wanted to measure something a bit a more subjective, such as happiness? If I'm looking for flu remedies, Google can draw a reasonable inference that I'm in the market for flu symptom remedies because I've got a bug. But as Paul points out, during the SARS outbreak millions of people began searching for information about SARS simply because they were curious about it.

Someone who types "I have the flu" is more likely to actually have it. To gain a sense of what people are feeling, what symptoms they may be experiencing, you need a platform that allows people to broadcast their current state of being and you need a user base interested in doing exactly that. This is precisely what Twitter is about. Paul and Dredze's system predicted not only flu but also revealed a wide assortment of illnesses and symptoms among the observed subjects.

Writing an algorithm that can parse nuances in human language is no easy matter. Humans are able to easily differentiate between words that clearly refer to illness, such as "101-degree fever," "nausea," and "seizure," from those terms that we *figuratively* attribute to infectious illness, such as "Bieber fever," being "bored to death," and "OMG that guy's Fuchsia Vneck is giving me a seizure #hipsterfail." We learn appropriate use of rhetorical, figurative, or simply nonliteral language based on a wide variety of feedback signals. A human's education in semantics, the study of the multiple meanings of language, is lifelong. Scientists hoping to imbue some shallow understanding of the slipperiness of words in computer programs don't have that luxury.

What Paul and Dredze did have was a ridiculous amount of data: a stockpile of 2 billion tweets that they built over a period of a year and a half. "Two billion over a year is a small sample, a lot of data, but a small sample," Paul explains.

Once they compiled their 2-billion-tweet corpus they soon found they had another problem. It was too big—not too large for a program but too large for them to work with. If the program was going to learn, they would have to design lessons for it; more specifically establish a set of rules that the program could use to separate health-related tweets from non-health-related ones. To write those rules, they needed to whittle the corpus down significantly.

The next step was to figure out which illness-related words would be the most fruitful. "Words like the 'flu' are strong. But names of really specific drugs return so few tweets they're not worth including," Paul says. They looked at which of the 2 billion tweets were related to thirty thousand key words from Web MD and WrongDiagnosis.com, indicating sickness. This filtering and then classifying gave them a pile of just 11 million tweets. These contained such words as "flu," "sick," and various other terms that were related to illness but often used for other purposes (e.g.,"Web design class gives me a huge headache every time").

These remaining 11 million tweets had to be classified or annotated by hand, a task that was monumental enough to be beyond the reach of a pair of linguists. But the emergence of crowd-sourcing platforms has reduced this sort of large-scale, highly redundant task to a chore that can be smashed up and instantaneously divided among thousands of people around the world through Amazon's

Mechanical Turk service. The cost, Paul recalls, was close to $100. Each of the 11 million tweets was labeled three times to guard against human error. If two of the Mechanical Turk labelers believed a somewhat ambiguous tweet was about health, the chances of the tweet's not being related to health was small.

The point of this exercise was not to relieve the program of the burden of having to learn for itself but to create a body of correct answers that the program could check itself against. In machine learning, this is called a training set. It's what the program uses to look up the answers to see if it's right or wrong.

Examples of machine learning used in practice go back to the 1950s, but only recently do we have enough material to train a computer model on virtually anything. This is a methodology breakthrough that is now possible only because of the Internet, where spontaneous data creation from users has taken the place of costly and laborious surveys.

What Paul and Dredze's program does is show health and flu trends in something closer to real time, telemetrically, rather than in future time. But remember, the future is a matter of perception, and perception on such matters as flu outbreak is shaped by reporting. Predicting what official CDC results will reveal two weeks before those results become public is an example of an area of a particular future's becoming more exposed where it had once been cloaked.

But let's return to our scenario in which Josh was told the identity of the person to whom he was going to give the flu. It also included a level of fine granularity, actionable intelligence on the likelihood, computed as a probability score, of direct person-to-person flu transmission. We aren't concerned with some big data problem, we're concerned about Josh!

A few years ago attempting to solve a problem so complex would have involved extremely expensive and elaborate socio-technical simulations carried out by people with degrees in nuclear physics. But that was before millions of people took up the habit of broadcasting their present physical condition, their location, and their plans, all at once.

Working off Paul and Dredze's research, Adam Sadilek published two papers in the spring of 2012 in which he showed how to use geo-tagged tweets to discern—in real time—which one of your friends has a cold, deduce where he got it, and predict the likelihood of his giving it to you.

He applied Paul and Dredze's program that separates sick tweets from benign tweets, on top of a real-world setting. Sadilek looked at 15 million tweets, about 4 million of which had been geo-tagged, from 632,611 unique Twitter users in the city of New York. About half of those tweets (2.5 million) were from people who posted geo-tagged tweets more than a hundred times per month, so-called geo-active users. There were only about 6,000 of these people, but there were 31,874 friendships between them. Each person had an average of about 5 friends in the group.

That gave Sadilek a window into where these 6,000 people were, what they felt while they were there, with whom they then met, and for how long. Based on that information Sadilek's algorithm allowed him to predict 18 percent of all the flu cases that were going to occur between these 6,000 people over the course of the next month. The experiment was a proof of concept to show the algorithm worked.[19] Had he and his colleagues utilized a larger scale, they would have succeeded in predicting more *individual* instances of flu transmission in one month than any epidemiologist in history.

Naturally, we can see the limitations of this exercise. Sadilek admits that in 80 percent of the instances when one of the subjects actually became sick with the flu, the causality was opaque. And he acknowledges his algorithm is entirely dependent on infectious people geo-tagging complaints about their maladies on a social network. The percentage of people in New York likely to sick-tweet, as in *Don't know which is more congested, this subway car or my nose #sickinuptheFtrain*, is extremely small. Geo-taggers represented 1 in 3,000 New Yorkers at the time of the experiment by Sadilek's calculation. Research from Brigham Young University and the Pew Research Center also shows that while more than 60 percent of Americans go online to look for health information, only about 15 percent actually post information about how they themselves are feeling, illnesses they've had, reactions to medications, and so on.[20] This is human nature: when we're sick, we don't share. And geo-taggers are a rare breed. These are people willing to release a ton of information about themselves, where they go, who they see, and how they feel. This isn't typical behavior . . . yet. But the number of people willing to make that sort of information public is much larger than it was three years ago and exponentially larger than it was ten years ago. This, in part, is why the Josh flu scenario is not far-fetched at all.

Breakthroughs such as these won't vanquish the flu. One persistent gap in coverage is flu transmission through surface contact (which accounts for 31 percent of flu transmissions). And these emerging capabilities also open up a new set of contentions and protocol issues. Take the simple school scenario outlined above. If you're a school administrator, you might respond very differently to a 10 percent chance of one child catching the flu than if you are that one child's parent. All administrators want to make the best possible choice for the greatest number of children under their supervision; all parents want to make the best decision for their own child. These motivations are sometimes in harmony but are often in conflict. If several children show up one morning with a high probability of giving a mild influenza to several other children, do you send the infected kids home? Do you alert the parents of the other children? Do you do nothing? If you're an employer and you know that several of your workers are going to come into the office with a mean bug, contaminate the office, and cause a lot productivity loss across your entire staff, do you instruct them—preemptively—to take a sick day? If you don't offer paid sick leave, do you change your policy or do you force the workers to take an unpaid day, effectively punishing them for being willing to show up for work while under the weather? What if the sick person isn't your employee but your office mate? How do you ask for time off because someone *else* has a cold?

In the past, the basic approach that supervisors passed on to their employees took the form of: Use your best judgment. People get sick, sometimes they spread their sickness and sometimes they don't. Deal with it. The ultimate costs of one person's cold, the number of people they'll infect, has historically been very hard to calculate. In the naked future, we may not have the luxury to feign that sort of ignorance. Indeed, we may know *exactly* how much a particular person's cold will cost the moment it shows up. It's a number that will change depending on the action that we—as managers, workers, students, and parents—take. The most effective solution for any individual won't be the best solution for someone else. These arguments won't be solved easily. But because we can see the future not just of flu, but of Josh's flu, of yours and mine, at least we can begin having the debate.

When we're sick we hide our weaknesses and inflate our strengths because instinct tells us to do so. No one wants to be treated differently or seen as ill. And so we regard with suspicion and amusement those people who will post evidence of their maladies to social networks, to out themselves as sick. We go through life privately feeling one way while showing a different face publicly and calling this behavior normal. We may soon realize that those people willing to share how they feel are committing a noble and selfless act. In some ways, they're healthier than the rest of us.

CHAPTER 4
Fixing the Weather

IT'S September 2011, and Wolfgang Wagner has just published his resignation as the editor of the journal *Remote Sensing*. To Wagner, this is a sad but simple matter. He has made a mistake. The journal had accepted an article arguing that because of previous errors in satellite modeling there was no clear way to tell if CO_2 was causing heat to become trapped in the atmosphere (a phenomenon called radiative forcing) or if the heat increases in the atmosphere were actually caused by clouds (radiative feedback).[1]

Though the article's author, University of Alabama researcher Roy Spencer, had done some important pioneering work in collecting temperature data from orbital satellites, he had repeatedly been forced to publish corrections for many of his major findings. As a result, Spencer was not a well-regarded researcher in the climate science community.

But Roy Spencer is no dummy.

He sought out *Remote Sensing* precisely because it was *not* a journal dedicated to climate change but to the study of satellites and the modeling of satellite data, a journal about *instruments*. His paper was a Trojan horse.

In his resignation letter, Wagner explains that Spencer's article hadn't been properly vetted and that it ignored other previous, contradictory findings. Wagner further says that in accepting the paper he had trusted Spencer's credentials as a former head of NASA's Microwave Sounding Unit, as well as the judgment of the journal's managing editor. He admits that he himself never really read the article before he decided to include it.

"Had I taken a look, I would have been able to see the problems right away," Wagner told me. Indeed, almost as soon as the paper was published, researchers from around the world were quick to call it seriously flawed.

By that point it was too late. Roy Spencer had scored a major political victory. The publication of Spencer in a "peer-reviewed" science publication was quickly picked up by conservative media. On April 6 meteorologist Joe Bastardi appeared on *Fox and Friends* to declare that global warming had effectively been debunked. A representative from the Heartland Institute writing on the *Forbes* blog said that the inclusion of the paper in a peer-reviewed journal ended the global warming debate once and for all.[2]

In the act of resigning, Wagner hoped to put an end to the controversy and also regain a normal teaching schedule. The job of editing an academic journal demands a great deal of work and the rewards are mostly personal at best. Wagner was not particularly attached to the title. He took up the position at the publisher's request and worked on a voluntary basis. "I had my own work to get back to," Wolfgang told me. "I was very busy at the time." His decision to remove himself from the masthead seemed an appropriate one in light of the error but not really a significant action.

Days later, Wagner's e-mail in-box was full of death threats. He describes these as "awful," mostly anonymous, and emanating primarily from the United States, Australia, and the United Kingdom. "I was glad I lived in Vienna," he says.

Spencer went on to claim on his blog that Wagner (whom he calls simply "the editor") had been forced to resign by the Intergovernmental Panel on Climate Change (IPCC). "It appears the IPCC gatekeepers have once again put pressure on a journal for daring to publish anything that might hurt the IPCC's politically immovable position that climate change is almost entirely human-caused. I can see no other explanation for an editor resigning in such a situation."[3]

Wagner insists that he's never met anyone from the IPCC, was never contacted by the organization, and certainly never received any pressure. He's a physicist and a teacher. His expertise is in detecting soil moisture using satellites (which, in practice, is even harder than it sounds). This is not to say he didn't make a mistake. He overextended himself and didn't appreciate that certain voices would rise to champion a discredited methodology as forcefully as one might defend family, country, or love of God. "It's very strange for us here. Climate change is not a controversial subject for us in Central Europe," says Wagner. "I was shocked."

Meteorological forecasting was not only the first big data problem but the challenge that actually gave rise to the computer as we understand it today. Therefore, the question of what we can infer about the future from vast amounts of computational data is tied to the problem of predicting the effects of climate change. It would seem that if science can't solve this problem, it should give up the entire endeavor of trying to predict anything but the results of lab experiments. So why, in spite of all our other successes science has made in big data aided computational prediction, are scientists such as Wagner still getting death threats in the mail? Why, indeed, are we still debating this stuff? Why can't we get it right once and for all?

The Memories of Tubes

The date is June 4, 1944. The setting is Southwick House, the tactical advance headquarters of the Allied forces, just outside Portsmouth, England. The day before was bright and clear, but clouds are moving down from Nova Scotia under the cover of night. The next month will bring with it uncharacteristically high winds and much April-like weather.

General Dwight Eisenhower had been preparing for the D-day invasion until just a few hours ago when his head of meteorology, British group captain James Stagg, informed him that the seas would be too rough for an invasion the next day. Eisenhower is now faced with the prospect of postponing the assault until June 19 and quite possibly losing the element of surprise.

At 4:15 A.M., Eisenhower assembles his top fifteen commanders. Stagg, this time, has better news. Having spoken with one of his meteorologists, a Norwegian forecaster named Sverre Petterssen, Stagg is now convinced the Allies will have a very brief window on June 6 to stage their assault on the beach at Normandy.

Petterssen was a devotee of the relatively new Bergen school of meteorology, which held that weather was influenced by masses of cold and warm air, or "fronts." These fronts collided miles above the earth's surface. According to the Bergen school, the fluid dynamics of these fronts, the density, the water content, the velocity of their movement, et cetera, when properly observed would yield a more accurate forecast than previous, intuitive, historical methods of forecasting (e.g., the simple statistical method that resulted in the original June 5 forecast). Petterssen calculated that a movement of the storm front eastward would result in a brief break in the wind and rain.

When Stagg informs Eisenhower of this, the general takes less than thirty seconds to reach a new course of action. "Okay," he says. "Let's go."[4]

James Fleming describes this decision in the *Proceedings of the International Commission on History of Meteorology* as a pivotal moment in the turning of the war:

> Ironically, the German meteorologists, aware of new storms moving in from the North Atlantic, had decided that the weather would be much too bad to permit an invasion attempt. The Germans were caught completely off guard. Their high command had relaxed and many officers were

on leave; their airplanes were grounded; their naval vessels absent.[5]

The lesson to future military leaders from Eisenhower's success was clear: better weather data, and better forecasting, were the difference between victory and defeat.

Skip ahead about a year. It is the fall of 1945 when Hungarian-born mathematician John von Neumann arrives at the office of Admiral Lewis Strauss in Washington, D.C. Von Neumann had been working on the development of the atomic bomb and was rapidly becoming one of the most influential technological minds in government. The point of the visit to Strauss was to request $200,000 for an extremely ambitious project, the construction of a machine capable of predicting the weather.

Accompanying von Neumann is Vladimir Zworykin, an engineer at RCA who had been instrumental in turning vacuum tubes into objects that could store information in an extremely compact form. These would be the essential components in the proposed weather prediction machine.

"It should be possible to assemble electronic memory tubes in a circuit so that enormous amounts of information could be contained," Zworykin told Strauss during the meeting, a feat that had already been demonstrated in a few prototype machines.[6] "This information would consist of observations on temperature, humidity, wind direction and force, barometric pressure, and many other meteorological facts at various points on the earth's surface and at selected elevations above it."

At the time, no computer was capable of holding enough information or executing the calculations necessary for weather prediction. Indeed, there was really no such thing as what we consider to be a computer at all. The closest thing was the U.S. Army's Electronic Numerical Integrator and Computer (ENIAC). The ENIAC's user interface was a constellation of dials that had to be reconfigured for every new problem. It took up a ridiculous 1,800 square feet of space, held 17,468 very breakable vacuum tubes, and required

160 kilowatts of electrical power to operate, enough juice to power about fifty houses.[7]

Von Neumann wanted to develop a machine that could compute a wider set of problems than the ENIAC and do so without programmers having to extensively rework it. This *automatic* computer would have a memory that could be accessed and used without a lot of changes, which he called the stored-program concept. The idea would later make possible the random access memory (RAM) functioning that is the very backbone of every home computer, every smartphone, every server, and the entire digital universe.

Von Neumann didn't anticipate that this idea would transform the world completely and totally. In 1945 he knew only that such a device would need to have some sort of military application if he was to get any money to build it. Weather forecasting lent itself perfectly to this because it had recently proven a decisive element in the D-day invasion.

But long-range weather prediction was not von Neumann's ultimate goal. What he was really after was a new kind of weapon, one of greater strategic advantage than any nuclear bomb. Predicting the weather, von Neumann believed, was the first step in *controlling* it.

The idea was spelled out in full in Zworykin's *Outline of Weather Proposal*, which stated: "The eventual goal to be attained is the international organization of means to study weather phenomena as global phenomena and to *channel the world's weather*, as far as possible, in such a way as to minimize the damage of catastrophic disturbances and otherwise to benefit the world to the greatest extent by improved climatic conditions where possible" (emphasis added).

In broaching this fantastical idea, Zworykin found an eager partner in von Neumann who wrote in his cover letter on the proposal (dated October 24, 1945), "I agree with you completely . . . this would provide a basis for scientific approach for influencing the weather." This control would be achieved by perfectly timed and calculated explosions of energy. "All stable processes we shall predict. All unstable processes we shall control." In weather control, meteorology had a new goal worthy of its greatest efforts.[8]

The meeting in Strauss's office was a success. And so in 1946 the Institute for Advanced Studies project was born with the financial help of the U.S. Navy and Air Force.

Two years would pass before von Neumann was able to complete his computer (a team from the University of Manchester would go on to create the first stored-program device far sooner). In the meantime, the institute used the ENIAC to make its weather calculations. The ENIAC was off-line as often as it was online and could only forecast at a very slow rate. If it was Tuesday and you wanted a forecast for the following day, you wouldn't get your forecast until Wednesday had arrived. But the meteorologists von Neumann brought to the project did have a remarkable and historic success; using the ENIAC they proved that it was indeed possible to predict the weather mechanically. The team was able to predict climate conditions three miles above the earth's surface with realistic accuracy. They soon turned their attention to the problem of improving the models and adding computational power to extend the forecast range and decrease the amount of time required to make projections.

For von Neumann, the progress was plodding. He wasn't satisfied with daily, weekly, or monthly forecasts. He wanted to be able to run a simulation or climate model and come up with a snapshot of the weather at *any* future time. This "infinite forecast" would reduce the stratospheric inner workings of air, water, and heat to discernible causal relationships, like the workings of a clock.

In 1954 he found encouragement in the work of Norman Phillips, a meteorologist who was experimenting with what could be called the first true climate model capable of making reasonable forecasts of troposphere activity thirty days into the future.

Von Neumann hosted a conference in Princeton, New Jersey, in October 1955 to discuss the importance of Phillips's work. This meeting morphed into perhaps the first major global warming summit, replete with all the discord later climate change conferences would hold. In *Turing's Cathedral*, British historian George Dyson's historical account of the field of computer science, Dyson recounts some of what went on:

> Consideration was given to the theory that the carbon dioxide content of the atmosphere has been increasing since the beginning of the industrial revolution, and that this increase has resulted in a warming of the atmosphere since that time . . . Von Neumann questioned the validity of this theory, stating that there is reason to believe that most of the industrial carbon dioxide introduced into the atmosphere must already have been absorbed by the ocean.[9]

It's the sort of discussion that could take place between a pair of very modern, twenty-first-century pundits, which evinces the fact that public certainty about climate change remains in a state of flux, even as scientific consensus has grown more firm over the last half century.

We've made great progress in understanding weather patterns and events and how they are influenced by climate, but we were never able to build von Neumann's weather clock, at least not the way he envisioned it. There is no model that perfectly mirrors the climate. The potential variables are too numerous for us to handle at present. The weather is a dynamic and chaotic system. Changes, imbalances, and disturbances that are too small to see build power quickly and meet with other physical forces, matter, and energy, creating new imbalances, adding new complications, and all of this is happening all the time, all over the world.

What we have instead of an infinite forecast is an amalgamation of timepieces, stopwatches, and calendars sharing many parts between them but purporting to tell time in somewhat different ways. Historical temperature readings from the 1800s provide one picture of our weather history (and thus our future weather); tree-ring samples going back centuries provide another; ice-core samples going back well before the dawn of humankind provide a third. Researchers who are spread all over the world, and who use a menagerie of instruments including radioiodine, offshore and surface weather stations, geostationary and polar orbiting data, all contribute to an ever-changing understanding of how the climate operates. All of these observational and mechanical processes yield hundreds of terabytes of data on a yearly basis. No wonder climate science as a field consumes more supercomputer time than even particle physics research. Standardizing and synthesizing all of this information is a monumental task.

So what do the models tell us? The IPCC has forecast a mean global temperature increase of 4 to 6 degrees Celsius and a sea level rise of as much as a meter by the year 2100. (Views from the 2013 assessment were not yet available at the time of writing.) This may not sound like a significant change, only a mere seasonal transition. We are again forced to rely on computational models to make sense of the transition, which suggest that such a rise, especially in the 4-to-6-degree realm, would be incredibly disruptive to human life on this planet. Cities such as Bangladesh and even the lower portion of Manhattan would be overtaken by water. Parts of the world that had enjoyed temperate or stable climate conditions would see drought, flooding, and an increased number of extremely strong hurricanes (Category 4 or higher).

The IPCC findings are not perfect and the body has been forced to issue corrections, amendments, and the like. Their assessments represent not what thousands of people at the forefront of climate science, academics, engineers, technicians, and politicians, *know* but what they can *agree on*. There is tremendous nuance within that consensus and some scientists believe the IPCC estimates are too conservative. Yet agreement on the existence of man-made climate change is far narrower than certain political and business interests would like to believe. More than 95 percent of climate scientists agree that climate change is happening and the process is heavily influenced by mankind.[10]

To this day, the fundamental question that drives the so-called climate debate is simply this: What can and cannot be modeled? It's a flimsy-sounding problem and so flimsy politicians who are eager to align themselves with the prevailing political winds make good use of its ambiguity. In 2011 presidential candidate and former House Speaker Newt Gingrich found himself having to reverse his previous support for carbon cap-and-trade legislation in order to appeal to a far more conservative Republican base voting in the Republican primaries. His explanation for this turnaround was laughable: "I think that we're a long way from being able to translate a computer program into actual science . . . I, also, am an amateur paleontologist, so I've spent a lot of time looking at the Earth's temperature . . . and I'm a lot harder to convince than just [someone who is] looking at a computer model."[11] Imagine all the trouble and argument that could have been avoided if John von Neumann had had the chance to meet Newt Gingrich. Perhaps he would have given up the whole computer nonsense and taken up a more suitable profession, such as paleontology!

It's very gratifying to shake a fist at know-nothings who claim that global warming is a grand hoax. But very little about the actual process of climate modeling is intuitive or straightforward. In elementary school we're taught that an experiment performed under a certain set of conditions should result in the same outcome no matter who you are or what you believe. We call this process science. But there is no way to run an experiment on the entire global climate. We're stuck with the numerical simulations that the average person can't repeat. Yet modeling is essential to make *any* sense of the data. It's not an after-process; it is *the* process. In the future, even as scientific consensus continues to cement, the politics of climate change will become more troubled and fractious; the assault on climate science will intensify.

Why is this? Public opinion on the existence of anthropogenic climate change moves up and down based on factors outside science. That's fairly incredible when you think about it: *more* certainty on the part of climate science results in *less* certainty on the part of the public. How could that be? The short answer is that climate science represents everything that's wrong with the big data present. The opaque and highly technical nature of crunching these various variables and data points assures that many people are left out of the process.

Most people without science degrees understand science from their elementary or high school exposure to the scientific method: you form a hypothesis, you devise an experiment, you record the results, you repeat. If anyone in the world can run your experiment and achieve the same results that you have, then you've learned something true. Of course, there are plenty of homemade science projects that enable students to understand the basic principles of climate change but a very small number of men and women actually have access to the data, the computer-processing power, and the technical skills necessary to develop climate models or critique them. Simply put, not every technically trained person has access to the rich data or the robust computing power necessary to make a climate model. When looked at from that (rather limited) perspective, climate science fails a crucial test; the models or experiments can't be simply or easily reproduced by, say, a local weatherperson. The perception among many is that we're being asked on faith to accept the research of high priests and clerics.

More important, the modelers can only agree on broad scenarios that look decades into the future. But they can't tell you how unmitigated climate change will affect anyone specifically or how it will change the weather in the next five or twenty days. We expect the smartest people in the world to be able to deliver extremely specific predictions, as in *climate change will cause a tornado this weekend; cancel your barbecue*. But that's not the way it works. In fact, climate scientists are, for the most part, extremely reluctant to blame increasing global temperatures on any specific weather event that we experience, even retrospectively. What they can offer is sound and sober admonishment to live more responsibly. The message of the climate science community too often seems to be that we should be prepared to make *individual* sacrifices or else face terrible *collective* consequences later. Is it any wonder why polling has shown that in tough economic times public skepticism toward climate change grows?[12]

For both Republicans and Democrats, developing nations and the industrialized world, attacking climate change is a loser at the ballot box and could soon become more so, especially in the United States. The same political forces that insist more research is needed before the United States can take meaningful action on climate change are working to undercut the nation's ability to conduct that very research. The United States is continually on the verge of ceding leadership on the most important issue of our species.

In 2013 NASA had thirteen Earth-monitoring satellites. Six will no longer be in operation by 2016. In 2006, under former president George W. Bush, the U.S. government restructured the National Polar-orbiting Operational Environmental Satellite System (NPOESS) and the Geostationary Operational Environmental Satellite-R Series (GOES-R), which worked to greatly diminish the United States' Earth-monitoring capabilities. A 2011 Government Accountability Report found that unless the United States works to cover the gap and put new equipment in space, the nation "will not be able to provide key environmental data that are important for sustaining climate and space weather measurements."[13]

In 2010 President Barack Obama warned that the budget put forward by House Republicans would seriously harm the ability of the nation to gather weather data: "Over time, our weather forecasts would become less accurate because we wouldn't be able to afford to launch new satellites and that means governors and mayors would have to wait longer to order evacuations in the event of a hurricane."[14]

That means future climate data will increasingly come from such other sources such as Europe, Russia, and even China. Much like global warming itself, this is a sort of feedback loop, which conveniently plays into the narrative that climate science is inherently dishonest, sneaky, and most—damningly globalist—un-American in nature. This will be at least partially true. The mounting evidence for man-made climate change will indeed *be less American* in origin.

Fifty years after von Neumann set out to change what could be known about the operations of our natural world, various political forces in the United States the adopted country he loved so dearly, are demanding we ignore the problem or employ only half measures to fix it.

What we need is a system or service to allow individuals to understand how climate change may affect them individually, not in the year 2100 but in telemetric real time (or something like it). Only then can the big data present give way to a future in which everyone can understand what climate change actually means *to them*. The very slender silver lining on the dark cloud of climate change is that entrepreneurs are stepping up to fill the role our public leaders have left vacant.

Your Climate Insurance Provider

The year is 2002. David Friedberg is driving down San Francisco's Embarcadero, the highway that hugs the San Francisco coastline, on his way to his finance job in Foster City. It's raining. At the intersection near the AT&T Park, Friedberg turns and sees the Bike Hut, a small plywood bicycle-rental stand on the boardwalk. It's closed.

"I said to myself, *That's a pretty crappy business.* Whether or not this guy is going to make money in a given month depends on how many days it rains. It occurred to me that's actually a big problem," Friedberg went on to tell a roomful of Stanford undergrads in a 2011 lecture.[15]

Back on the Embarcadero, Friedberg realizes that if he can model the weather, with a sense of the probability of different weather outcomes, he could provide businesses such as the bike shop with individual insurance against weather loss. The owners of the Bike Hut could buy an insurance contract to pay him a small sum when it rains. Friedberg stores this insight. The light turns green. He continues his commute to Foster City.

Friedberg, born in South Africa, came to the United States in 1986 at the age of six. His parents moved the family to Los Angeles hoping to succeed as independent filmmakers. In his Stanford lecture, he recounts growing up in Hollywood and watching his parents struggle to find investors for their projects. The experience might have scared him away from risk, long-shot bets, and overly ambitious plans. Instead, he says, his parents instilled in him the exact opposite attitude. As a child he had been interested in "the way things work" and went on to study astrophysics. But the study of the interrelationship of solar bodies, while challenging intellectually, felt divorced from real life. He felt like "a cog in a machine that was going to take forever to output some theory that might be disproven fifty years later." At this time, around 1999, the tech bubble was reaching critical mass and Friedberg acknowledges that the hype helped move him toward investment banking in Silicon Valley.

A few years later he moved from finance to Google where he worked as a product manager in the AdWords program. This involved sifting through massive amounts of data, anonymized versions of the words and phrases that Google users were putting in e-mails and search queries. His mind kept returning to the shuttered Bike Hut and how that business model could be improved. He saw a clear overlap. "The idea of taking lots of data and extracting signals could apply to weather," he would go on to tell the Stanford students. "We could analyze weather data and take these signals and apply them to a problem." That problem was two-pronged: what the weather would be and how much it would cost. If you could calculate those two things, then you could sell an insurance contract to a ski resort to help them recoup losses after a season with below-average snowfall, or you could help a Bike Hut make back a bit of lost profit after a very rainy spring.

He solicited the help of fellow Googler Siraj Khaliq as chief technical officer and, with $300,000 in seed money, in 2006 they formed WeatherBill (since renamed the Climate Corporation). The first step was to get a clear statistical sense of the different types of weather a given location was most likely to experience in the future based on what had happened in the past and what was happening right now.

They gained access to live weather feeds through the National Oceanic and Atmospheric Administration (NOAA) and the National Climatic Data Center (NCDC). Khaliq tapped his network at Stanford and Friedberg looked into climate data sources from Berkeley. The data had to be up to the minute and constantly streaming;

telemetric, not numbers on tables. And there had to be *a lot* of it. They began aggregating data going back three decades from two hundred weather stations, and quickly doubled that number. Ultimately the product they would be selling was a climate model, just like those pesky IPCC models that prove so inconvenient for politicians. But these models would tell you the probability of a specific place getting too much sun, too much rain, and the like within a given time that was of relevance to a stakeholder. That's a very different product than a big projection about the global climate a hundred years hence.

But which industries were the most vulnerable to financial loss from weather events? Neighborhood bike shops, it turned out, didn't constitute a large enough market on which to found an insurance business. On the other end of the spectrum they considered the travel industry, but this field was dominated by large airlines that could afford to self-insure. Energy was another option, but that market had too few players as well. Friedberg and Khaliq soon realized that what they were selling wasn't exactly new; it had existed prior to 2007 in the form of the infamous derivative contract.

Unlike the rest of the derivative market, which has become synonymous with shady and complicated backroom deals that produce money but not value, derivative contracts in the context of weather insurance are actually straightforward. Let's say you are a big energy company and you do deep-water drilling off the coast of Mexico. You have offshore rigs that are vulnerable to big weather incidents. Massive storms can cause production delays, extensive damage to equipment, and worse. To insure—or "hedge"—against financial loss from a big weather event, you can go to a financial services firm like Goldman Sachs and ask to buy insurance. Goldman Sachs then does what Friedberg was trying to do: create a working model to figure out what sort of weather events were most likely and, from that, how much it would cost to insure against weather loss. The way Goldman Sachs does it is extremely expensive; they build these models for a very small customer base. For Friedberg that was not an option. Agriculture was the largest, most underserved market. Individual farmers needed a way to insure against crop loss but couldn't afford the hedges Goldman Sachs was selling in the form of derivative contracts.

Here then was the opportunity to do the same thing big Wall Street banks charged huge service fees for, but Friedberg was going to do it for millions of people on demand (at scale), automatically, and at a much lower cost, effectively scooping up the business of small farmers that such companies as Goldman Sachs didn't care to reach. That they had no experience in agriculture (or insurance, for that matter) didn't seem to matter.

They could limit the variables the system had to consider by focusing on specific crops such as corn, soybeans, and wheat. Each crop needed a certain amount of sun, water, and growing time, and could fetch a particular price at market. They next had to figure out what the weather parameters, thus the potential loss, were going to be for the farmer growing that crop. "We realized we had to measure the weather much more locally. No one wanted coverage for weather a hundred and fifty miles away," Friedberg told the Stanford students.

They expanded their weather data sources to millions of daily measurement sources. Currently they use that data to build simulations for each potential location. They run the simulation ten thousand times, a process that takes several days. The company generates 25 terabytes of data every two weeks. If you ran that simulation on a conventional server, it would take forty years. If you ran it on the ENIAC, it would take centuries. Today, they farm the heavy processing to the cloud, specifically Amazon's cloud services. They are now one of the largest users of MapReduce.

The result: the Climate Corporation can output the distribution of possible weather outcomes for any 2.5-by-2.5-mile area in the United States, on any given day, *in the next two years* (730 days). "It is the probabilistic distribution of things that might happen that allows us to figure out what price to charge for the insurance that we are selling," Friedberg told MSNBC reporter John Roach.[16]

In other words, they're assessing the risk of financial loss from weather, but they're doing so with a level of precision that surpasses what your local weatherperson can do. They're not predicting the weather; they're predicting it ten thousand times per customer, and learning more and more about the sort of weather that particular customer will experience every time they run that simulation, in much the same way that if you were to sit down and watch ten thousand games of tennis, your ability to predict the next Wimbledon champion would be better than average. It's not magic. It's a statistical trick with an enormous number of data points, including not only what's happening in the clouds but what's happening in the soil, what stage of growth the crops are in, and what seed a farmer planted.[17]

Are the estimates good? Ultimately, the market will be the judge. The government already offers farmers some coverage against loss. Climate Corporation's Total Weather Insurance (TWI) is designed to complement what the government does already. Farmers don't tend to have a lot of discretionary income for supplemental insurance, so if the company's product is too expensive, it will lose customers fast. Tellingly, TWI has become one of the most expensive insurance products on the market for farmers. And as drought conditions worsen year after year, the price is going up. In 2012, throughout parts of Kansas that were experiencing extreme drought, Climate Corporation charged $30 to cover an acre that would only yield $50 in profit.

Climate Corporation doesn't have to put any of its or its backers' money down to settle claims. The money the firm pays out comes from reinsurance, which, in effect, is insurance for insurance companies. The private reinsurers that the company deals with are multibillion-dollar outfits; they won't go belly-up based on claims from a few unlucky winter wheat growers, but that doesn't mean that those contracts, if they're poorly designed, won't lead reinsurers to raise prices on Climate Corporation, which in turn will force Climate Corporation to raise premiums on its customers.

Part of the reason for the already big bill is that the company sends a lot of money back to its policyholders. In 2012, 80 percent of its policyholders who were corn farmers received a payout from Climate Corporation for crop loss. In the states of Illinois, Nebraska,

Colorado, Kentucky, Missouri, Oklahoma, and Tennessee, where the 2012 drought hit particularly hard, virtually every Climate Corporation customer got a check.[18]

More remarkable is how these checks are issued. When the company detects that water, heat, or other conditions have reached a point where the crop will suffer damages, the system calculates the cost and sends the payout. The policyholder doesn't lift a finger. This is Climate Corporation's edge: it pays off far faster than the government insurance it's supplementing.

It sounds like a business model that could never endure. Yet it's booming precisely because the company is able to adapt the price of a policy to reflect the distribution of loss risk. Friedberg claims that far from damaging his business, the major drought of 2012 fell within the distribution of outcomes their models projected. The company is increasing the number of counties it covers by about a factor of two per year. In 2013 it doubled its number of agents from the prior year. Climate Corporation believes it is first in line to protect the $3.8 trillion agriculture market.

Is Climate Corporation an infinite forecast machine? No. The weather is growing less predictable, but our modeling abilities are advancing more quickly. In this way, Climate Corporation is much closer to von Neumann's dream of influencing the weather than is the IPCC. Whereas the importance of the IPCC is waning precisely because it can offer nothing but a number of scenarios, Climate Corporation has learned how to make money by predicting what the weather will cost. Unless you're fighting a war, knowing what the weather will cost you is as valuable as knowing what the weather conditions will be. That's the difference between the big data present and the naked future where people have telemetric data to make individual decisions about what to do next. Are we any closer to controlling the weather? The answer is both yes and no. The ability to mitigate risk is a form of control.

It is worth noting that Climate Corporation was recently bought by Monsanto, the controversial company most closely associated with genetically modified foods and for a number of patent lawsuits with farmers. Monsanto may use Climate Corporation's data to engineer new, genetically novel seeds that are more resistant to heat and water stress, which could be a boon to the fight against global hunger. But not all of Monsanto's business practices are in line with the public interest and they may take the same protective approach to climate data as they have taken to seeds, restricting access to this important resource.

Because we live in the age of the vaunted entrepreneur, when even our most nominally right-leaning politicians make frequent habit of praising the free market and all its wondrous efficiencies while denigrating government as bloated and inefficient, we may draw from the story of Freidberg and von Neumann the simple yet wrong conclusion that business was able to adapt to our rapidly changing climate where government failed to arrest it because the business mentality is inherently superior to that of the public servant.

We would do well to remember that Climate Corporation does not have to take a direct stance on man-made climate change to sell its product. It's a company that provides a real and valuable service but it isn't fixing climate change so much as profiting from a more advanced understanding of it. Higher corn prices and decreased crops can create profit but they do not—in themselves—create value. Ultimately, we will have to fix this problem, and government will have to be part of that solution.[19]

If not for the wisdom, creativity, and genius of people who weren't afraid to be labeled public servants, there would be no international satellite data for NOAA to help Climate Corporation improve its models. There may not even be computers, as we understand them, on which to write code or do calculations.

Today, in many respects, we are moving backward on climate change even as we have learned to profit by it. But we are finally just beginning to understand what climate change means to us as individuals, which, perhaps ironically, could be the critical step in addressing the greatest problem we have ever faced. As the big data present becomes the naked future, we may still be able to save our species as well as many others from the worst consequences of our excess.

CHAPTER 5
Unities of Time and Space

THE date is February 29, 2012. The setting is the O'Reilly Strata Conference in Santa Clara, California. Xavier Amatriain, engineering manager of Netflix, is concluding his presentation on how the company recommends movies to users. In 2009 Netflix launched a $1 million prize to build a better recommendation engine. The conditions for the award were: the winning algorithm had to correctly predict the next 2.8 million ratings, around 6 per user, 10 percent more accurately than the current Netflix system (10 percent defined by rote mean-square deviation).

Not a single entry effectively hit the 10 percent mark but Netflix awarded a team called Bellkor the prize for a composite algorithm that performed 8.8 percent better than the system it was using at the time. Netflix has since stopped using it.

I am in the audience for this presentation. Like a lot of people, I use Netflix begrudgingly as I rarely, if ever, like the movies the site recommends for me. I have no attachment to Gritty Sci-Fi Fantasy, Dark Independent Thrillers, Heartfelt Movies, or Controversial Documentaries. I don't like movies because they share some cobbled-together thematic resemblance to a movie I just watched. I enjoy the cinema offerings that I enjoy because, quite simply, they're good. I can predict that I'm going to like a particular movie almost as soon as I begin watching it.

Today, Netflix has more than 5 billion ratings and receives 4 million new ratings every day. It streams billions of hours of content every three months. Millions of users now stream movies from services like Netflix. In so doing, these users create unique telemetric data about ratings and even at what point people start and stop watching particular movies—data that go toward revealing not only how Netflix users on average react to different scripts but also how individual tastes change from viewer to viewer and movie to movie. What I want to know from Amatriain is: Why can't Netflix predict I'll like a movie any better than I can?

I approach the microphone.

"In that ninety-nine percent of movies are crap, shouldn't there be more of a research effort to separate movies that are actually good from movies that are bad—research beyond genre and adjective tagging? And aren't you the best person in the world to lead that effort?" I ask.

He stares at me coldly. "I would argue that the concept of quality is very subjective," he answers. He goes on to explain that because of this subjectivity there is no way to predict what any individual person will actually like in a movie. The most we can hope for is a system to predict the *types* of movies someone *might* like.

If we assume the human response to art lends itself to some sort of analysis, then what precisely can be measured and how?

At the same time that I was having this conversation with Amatriain, Netflix was in the process of doing exactly what I was daring it to do: using its storehouse of user-generated telemetric data to predict and respond to its viewers. In February 2013 Netflix debuted its second original series, a political-suspense drama called *House of Cards* that was based on an old British television show. The new piece was set in Washington, D.C., that hotbed of Macbethian intrigue. *House of Cards* is an example of what will probably be called "optimized television." It represents all the important bits of information Netflix has been able to glean from silently observing the digital-viewing habits of its customers.

As Andrew Leonard noted in his write-up of the show on Salon.com, "Netflix's data indicated that the same subscribers who loved the original BBC production also gobbled down movies starring Kevin Spacey or directed by David Fincher. Therefore, concluded Netflix executives, a remake of the BBC drama with Spacey and Fincher attached was a no-brainer, to the point that the company committed $100 million for two 13-episode seasons."[1]

Netflix, like Amazon, knows that correlations across data sets don't offer scientific certainty, but they are enough for sales. Netflix isn't out to answer *why* people like to binge on political-suspense TV, why Kevin Spacey appeals to those audiences, or, indeed, why certain plots keep interest better than do other plots. The question of what causes a film or television drama to be good can be left to the art majors; Netflix has got subscriptions to sell. Every time we see a notice saying "People who bought this item also bought . . ." and we succumb to the urge to follow the herd and click "Buy," we show that this strategy works.

Back to the question of what can and cannot be predicted about individual taste from telemetric data. There are two ways of considering this problem. The first approach is Amatriain's, who said, "The concept of quality is very subjective." According to this line of thinking, with rankings, recorded viewings, and friend recommendations, you can make a number of determinations about the type of movie or television millions of people will watch, how they will watch it, and how to keep them watching. But an algorithm can't begin to figure out *why* a person likes the movies that she likes.

The second line of thinking can be summed up thusly: all of the above is bullshit.

The Engineering Professor

The year is 1996. University of Pennsylvania marketing professor Jehoshua Eliashberg is at a movie multiplex in downtown Philadelphia. He finds all the movies on offer unappealing. This is hardly an unusual experience, but Jehoshua Eliashberg is a rather unusual man.

Born in Israel, the son of a prosperous executive, Eliashberg became obsessed with logical problems at a young age. He studied electrical engineering at the Technion–Israel Institute of Technology but longed for a human challenge; he went into marketing and excelled, eventually landing a permanent faculty position at Wharton, consistently ranked one of the most prestigious business schools in the world.

Standing outside that movie theater in 1996, it occurred to him that there had to be a more scientific way to help the theater manager pick better movies. Eliashberg did some digging and discovered that all the decisions about what would show at his local theater came out of a central office in New York. In fact, the head of this office made all the booking decisions for theaters in this multiplex-movie chain across multiple states. This was the person who was going to pick what would play in Eliashberg's local theater. From a management perspective, this seemed terribly removed. Obviously, the guy was not in close touch with customers.

Eliashberg concluded that to make better decisions about what movies to book and how long to book them, the regional manager would need to know an estimate of three things: the national box-office revenues for particular movies, the expected revenues for individual screens, and the box-office revenue for particular films on a weekly basis, as a typical movie run was less than fifteen weeks, but could vary.[2]

The model that he and his colleagues Berend Wierenga and Chuck Weinberg went on to publish was useful for multiplex-movie managers who already had three weeks of box-office data on a movie they were showing and were looking to decide whether to let the movie continue to play at the later stages of the life cycle or to replace it with something else. It was a good result. But he wasn't satisfied. There had to be a way to apply some scientific knowledge to the actual *making* of the movie and improve not just the array of products on offer but do so before millions of dollars had been lost in production through an improved green-lighting process. There had to be a way to use statistics to make better films.

The first step was to establish the parameters of a problem: What were the components of a good movie? Here is where a scientist interested in only a pure quantification experiment would have thrown up his hands, as taste is subjective after all. But Eliashberg believes there are certain universal truths to storytelling and that these truths are accessible. He and his colleagues began reading the works of prominent film critics, including Irwin Blacker's *Elements of Screenwriting*, Syd Field's *Screenplay: The Foundations of Screenwriting*, and James Monaco's classic *How to Read a Film*.

In so doing, they took what some scientists might consider to be too great a liberty by basing their theory on the writings of experts, people who had deep experience in the areas of theater, film, and storytelling. The model would be built around their (gasp!) *opinions*. Eliashberg wasn't interested in money balling the movies. He didn't want to throw out the insight that had never been quantified. He just wanted to understand how different expert insight could fit into a working model.

"I'm always looking to solve problems in a more rigorous way," he explained to me, his voice anchored by his deep Israeli accent. "However, my experience has taught me that the best solution is usually a combination of intuition and science, human decision making guided by a formal model. Scientific methodology by itself doesn't do very well, but human intuition by itself also doesn't do well."

Many of the critical opinions were related to some best practices for narrative, how to build tension, craft sympathetic characters, create drama, time a joke, and so on. Other suggestions were specific to an uncanny degree: How long should a flashback be in an action movie? Does a topical and important premise play better in a suspense film than in a comedy? How should dialogue be distributed among the characters in a family film, a political thriller, a horror movie? How many scenes should be set inside or outside ("interior" versus "exterior" in script parlance)? What is the optimal title length? One of the most important considerations is genre. Some genres consistently perform better than others but any film can perform well if it hews close to those combinations of elements that are key to how its genre functions.

This idea actually forms the basis for all film criticism, but it predates Monaco and any other critic. It was invented before film and even before what we today call theater. It comes from a time when performance was understood to be the recitation of poetry.

It was Aristotle who first established the idea that different genres may have different rules. In his seminal work on art, *Poetics*, Aristotle singles out three genres of poetry: comedies, tragedies, and epics. Epics were the blockbusters of the Hellenistic period. They involved flights of fancy, special effects, and the impossible. Comedies were works about stupid, base, or low men. The most important genre for Aristotle was tragedy, which he defined not as we do today—stories with sad endings possibly involving teenagers who kill themselves—but as "an imitation of an action that is serious, complete, and of a certain magnitude." Aristotle lays down a series of guidelines for how to craft these certain-magnitude stories. The most important of these insights are the three *unities*:

Unity of place: The action should be confined to a specific locale, a town, a castle, et cetera.

Unity of time: One revolution of the sun or a single day is the optimal time span for a dramatic work.

Unity of action: Basically, all action should move the plot forward in some way. Cut out gratuitous action if at all possible. This rule is the most closely related to the *certain-magnitude* portion of the description of tragedy.[3]

Carl Foreman's script for the 1952 Western *High Noon* (for which Foreman was nominated for Best Original Screenplay) observes all three unities perfectly. The stage is set in the one-road town of Hadleyville in New Mexico Territory. It opens with newly retired marshal Will Kane marrying Amy Fowler, a beautiful young Quaker woman played by Grace Kelly. When Kane learns that his nemesis, the outlaw Frank Miller, is due in Hadleyville on the noon train accompanied by three roughnecks, Kane attempts to rally the town to mount a defense. But he's abandoned and betrayed by the people he serves at every turn. In the span of a single morning, Kane goes from happy newlywed to town pariah. No one expects him to survive the day. The action culminates in a gunfight. Kane emerges victorious, but he and his young wife have been transformed by the ordeal, and the people of Hadleyville for whom Kane has sacrificed all to protect have been revealed as cowards. The movie ends before the sun sets on the day.

Aristotle saw very different rules for each of the ancient genres. An epic was successful if it delighted the audience. Epics were often about the past, episodes that could be embellished for the glory of the historical figures depicted therein. The action could be random and disconnected so long as the ultimate effect was maximum entertainment. Today, *Pirates of the Caribbean* and the *Star Wars* franchise could all be considered Aristotelian epics and likely would not be improved by any observance of the unities.

Action in tragedy, according to Aristotle, should be "comprised within such limits, that the sequence of events, according to the law of probability or necessity, will admit of *a change* from bad fortune to good, or from good fortune to bad" (emphasis added). This is an important point that strikes directly at what Aristotle saw as the purpose of poetry and of storytelling. A work of tragedy presents a hypothesis about causality. As a method for exploring consequences it asks: How does the world work? How do humans operate within it? Why does misfortune befall the good? The credibility of the storyteller is supremely important in broaching these weighty subjects. Giant leaps across time and space damage the inherent believability of a narrative, even if these jumps are entertaining.

A theatrical work of any genre, whether comic, epic, or tragic, can be successful so long as its author understands its true nature, according to Aristotle. This is why there is such a thing as art that is good and art that is bad, quantifiably. One consumer may prefer an artistic-suspense piece to a family-adventure flick but surely everyone prefers a *good* family adventure to a *bad* suspense film. More important, this system can be used to calculate how much money a particular script will make when produced.

In a 2010 paper titled "Green-lighting Movie Scripts: Revenue Forecasting and Risk Management," Eliashberg puts forward a statistical method for forecasting a movie's box-office success on the basis of whether it's a good example of its genre type. He calls his system a Bayesian Additive Regression Tree for Quasi-Linear (BART-QL) model. Rather than give a single-point forecast, his system provides a range of forecasts that change when new information is added (and naturally it also provides a confidence interval).

To create the model, Eliashberg and his coauthors Sam K. Hui and John Zhang amassed a data set of two hundred movie scripts for films that were released during a ten-year time frame (1995 to 2006). For each script, they recorded the genre, how much the movie cost to make, and how much the investment returned (domestic box-office revenue). They then had a group of readers score the movie on the basis of twenty-three "content" parameters that came straight from the critics. Did the hero have a strong nemesis? Did the characters evolve as the movie progressed? Was the ending surprising? Did tension build? Was the premise believable, important, or both?

They also looked at how elements in the script interacted by analyzing how the action was described according to a list of thirty weighted key words. "Ship," "sword," and "chamber" received scores of -.25, -.35, and -.29, respectively; "girl," "mom," and "office" got .24, .24, and .16. Eliashberg terms these "correlation coefficients," meaning that words with similar weights are more likely to show up in the presence of one another if the movie is written in a way that's genre consistent. A negative score for many of the key words does not in any way mean that your movie will tank. You just want your script populated with lots of key words that have similar values. You want swords and chambers together, not swords and offices. But there's room for nuance. The BART-QL method can then determine which combination of variables were most likely to produce success and by how much.

The result is an algorithm that weights the wisdom of the critics and can tell you how well your script will perform at the box office. The primary variable, not surprisingly, is initial spend. Movie production budget and box-office take have a 70 percent correlation, which suggests that movie studios are already fairly good at allocating resources to the right script, and resource allocation in the movie business is *somewhat* of a self-fulfilling prophecy. But the effect diminishes after a certain budget point. In those movies in which the correlation coefficients are tightly packed together, where the men on ships with their swords stay away from the moms in the offices—movies, in other words, that exemplify their genre—the likelihood for success is greater.

The BART-QL can tell you how much a movie will make with a degree of accuracy measured by a mean-square error of .8698, versus a mean-square error of .9406 for a linear regression (average) of movie budgets alone. That may not sound like much but it works out to millions of dollars per movie.

More important, once you tinker with the script, the model can tell you if you've made the movie more profitable or less, by how much, and again with what degree of confidence. In this way it gives you a naked future view of how a potential audience will respond to changes that you make. The approach doesn't impose strict taboos. It won't tell you what sort of script to write. Rather, it provides a means to quantify the *cost* of certain decisions in plot or character over others. If you really need that scene where the pirate (brandishing his sword) gets sucked into a time warp and finds himself in the middle of a busy insurance office, you can keep it, but that decision really limits the other choices you can make. It's a high-cost jump that will tax the audience's patience and use up goodwill.

"Everything is conditional," says Eliashberg. "I can't tell you that 'you want to maximize the number of exterior scenes' or 'try to set the dialogue such that every character has the same amount of air time.' It all depends on many things. This is why I'm not in a position to recommend an optimal story line."

In the BART-QL method, the wisdom of the critics, and the great Aristotelian tradition they represent, has been codified and systematized. Whereas the human critic sitting in the darkened theater allows the experience of film to wash over him and then recalls bits and pieces later to construct a review, the formula recalls the most essential ingredients and then calculates how they interact with one another.

There's nothing in this that's beyond the capability of David Denby or A. O. Scott. The difference is a question of effort and time spent. Just as it takes an extremely gifted and ambitious computer scientist to undertake the sort of personal record keeping and self-quantification displayed by Stephen Wolfram and Ray Kurzweil in the 1980s, only the most eccentric critic would try to review a film by recording every plot point, the number of times particular words appear, the length of external scenes, the distribution of speaking time across the players, et cetera, and then pronounce the movie a success or a failure. But the same price-depreciation phenomenon that's enabling more people to collect and use personal information with less effort is playing out in the way the Eliashberg method serves as a critic with a perfect memory. Instead of giving consumers advice on which movies to see, it tells studio heads which scripts will earn how much and how to make scripts better. In so doing, the method looks toward a different series of hypothetical futures. One day it may be available in the form of an app that screenwriters keep on their phones.

Studios develop anywhere from twelve to twenty-five movies a year, but they keep one hundred to four hundred film scripts in development. That's a lot of money and energy going into making films that don't make back what it cost to produce or market them. The main problem with the studio system, says Eliashberg, is that big choices are made in a relative vacuum of information.

"They [studio executives] sit down in a green-lighting meeting, look at a script, and come up with a statement like: this is *Spider-Man* meets *Superman*. Then they look up *Spider-Man* and they look up *Superman* and they see how they performed in the box office. They take an average. That's how they decide the fate of a new script."

Eliashberg calls this "high-level comping" because this process examines the script from a height that omits important detail and compares it with another script from the same thousand-feet-up vantage point. It does nothing to solve the question of why so many people like *Spider-Man* enough to see multiple movies about him but so few people like *Green Lantern*, *Punisher*, or any of the other recent superhero super flops.

Although Eliashberg's formula could save studios millions a year, it hasn't been widely adopted. Like any true talent, he feels unfairly slighted by the Hollywood establishment and all the big egos therein. "You say, 'we developed a methodology to help you more efficiently green-light scripts.' They say, 'I've been in this business forty years . . .'" Still, he's optimistic. More and more, Hollywood is bankrolled by hedge funds and more and more hedge funds are run by Wharton students of the sort that show up in his class every semester.

But even if you can solve the question of which movie to make for maximum box-office spend, you may be working out the wrong equations; box office is not going to be the determinant of success that it once was, as movie-theater attendance has been falling for more than a decade.[4] This shouldn't surprise you. Today's movie-going experience—modern though it appears and feels—is an artifact of the industrial age. Millions of people pouring into cities to work in factories provided the necessary market for a new form of entertainment, one that was based around what was a very new technology. The device revolution has given us the movie theater you carry in your pocket. The allure of the velvet-seated film house is on the decline. The greater mystery is why it survived for so long. The rise of such streaming services as Netflix suggests a future of more personalized movie viewing and a *relative* waning of blockbusters in the future.

Herein lies my major disagreement with Netflix's Amatriain. Netflix, at the moment at least, seems willing to perpetuate the studio system's worst practices. In fact, high-level comping is exactly how Netflix recommends movies to you, on the basis of genres and actors in which you've expressed interest and how your interests correlate with others. The problem is that these genre tags and *House of Cards* tricks are only an approximation of the story itself. Netflix is firmly stuck in the big data present.

Bayesian systems such as the BART-QL could be most useful in extrapolating from *sparse* data sets, such as the sorts of movies one viewer has seen and liked. "Let's say you have multiple consumers, and for each consumer you have very few observations, like what movies he or she has seen, or liked, or whatever. Bayesian analysis allows you to aggregate across different consumers and make individual level predictions for each consumer, in instances where for each consumer you have very few data points," says Eliashberg.

There is no perfect movie but, in the naked future, there may be a statistically perfect movie for *you*. Locked away in Netflix's servers is, perhaps, the data to answer the question of why you like the stories that you do.

What would this look like? The perfectly personalized film probably has a lot in common with a very modern, extremely story-intensive video game in that the viewer (or player) would be able to make numerous decisions about the characters populating the film and even provide input into the story. That level of influence could be customized for different users. Because more and more video games have an online component, they provide a unique opportunity for telemetric data collection and thus prediction. In 2011 a group of researchers from the University of North Carolina showed that they could accurately predict how a player would respond to a key challenge or threat in a massively multiplayer online role-playing game, data that game developers could use to make their games better, both in the design stage and later, during game play.[5]

Imagine that same data collection applied to the act of selecting a movie to watch from a streaming service such as Netflix. Let's say you like thrillers that end on a down note but where the protagonist lives. That system should be able to steer you toward movies that meet that plot criteria. What about surprise twists? You probably have a tolerance for certain types of surprises over others (there are surely some that you hate). At the very least, a system looking to anticipate your absolutely perfect Friday 9 P.M. movie should be able to keep you from having to see the love interest die, or the charming nerd character become a tough guy, or Kevin Costner's bare ass if that's the sort of thing that a system with enough data on you can predict will ruin your night.

Video games, by definition, are an active experience. We think of movies as primarily passive, the same way we think of reading. But in the naked future there is no such thing as a purely passive entertainment experience. For instance, as millions of readers move from print to e-readers, both book retailers and publishers are discovering new insights about what makes a good book.

In July 2012, *Wall Street Journal* writer Alexandra Alter broke the story that Amazon, through its digital-book sales, had already begun telemetrically analyzing reader behavior to better predict which books will sell better overall, and which books will sell better to individuals. The rise of e-books is creating digital, data-driven dialogues between readers and publishers, and one day, maybe even between readers and authors, transforming, in Alter's words, "the activity [of reading] into something measurable and quasi public."

So far, Amazon has discovered that people who read popular series books like Harry Potter tend to read them rapidly and completely, that serious nonfiction gets read in spasms of interest, and that literary types, who you might think would be the most patient and determined readers, are the most likely to jump from book to book, passage to passage, like small-winged birds from branch to branch. If you have the Kindle app installed on your Mac, as I do, you can actually see which passages of books others have highlighted and how the world is responding to the text in real time. (By the way, please highlight that. Thanks.)

"The bigger trend we're trying to unearth is where are those drop-offs in certain kinds of books, and what can we do with publishers to prevent that?" Jim Hilt, Barnes & Noble's vice president of e-books, told Alter, "If we can help authors create even better books than they create today, it's a win for everybody."[6]

Not every consumer is comfortable with the notion of their e-reader's reading them, or with Netflix's analyzing not just personal movie choices but also movie-watching behavior in order to deliver more personalized recommendations. True, back in the 1990s (hardly a golden age for personal privacy) people felt the same way about Amazon features that are common today, such as the notification appearing at the bottom of the Amazon screen that says "Customers Who Bought This Item Also Bought . . ." "People were freaked out about that in 1998," Russ Grandinetti, vice president for Kindle Content, told a group in San Francisco. Today, says Grandinetti, "We've seen higher rates across the industry of people willing to share." That's good news whether you want Amazon tracking your behavior or not. Even if some people opt out, they'll still enjoy better recommendations because millions of other people let Netflix watch them watch movies.

Even if the entertainment products themselves don't seem to be changing, best-practice guidelines for creating entertainment are in some surprising ways. In 2010 psychologists James E. Cutting, Jordan E. DeLong, and Christine E. Nothelfer published a paper showing that, over the course of the last seventy years, the length of shots in particular movies was coming ever closer to resembling the attention patterns of the human brain, as measured via a mathematical wave analysis technique called the Fourier analysis. Movie shots, overall, are getting shorter. The average shot in the 2008 film *Quantum of Solace* lasts 1.7 seconds. Compare that with Alfred Hitchcock's experimental 1948 film *Rope*, which includes several shots that last nearly 10 minutes.[7] But shot length won't continue to collapse until it reaches nothing. No one wants to watch a movie composed of millions of millisecond flashes of story. So what's the optimal shot length under what circumstances? By bringing the current trend in line with a larger theory of human attention, Cutting's research provides a framework to predict what movies of the future will look like: they'll resemble, at least in feel and rhythm, the chaotic world that made us the animals we are today.

Ultimately, the real benefit of telemetrically tracking how people consume artistic works will be artists who will have new tools to determine how the pieces of a story might fit together for maximum effect. Some might argue that quantitative analysis is a poor substitute for real creativity and they are absolutely right. But let's be honest, our anxiety about this possibility arises from the notion that our innermost thoughts, feelings, and passions—the very human emotions we bring to the experience of viewing art—may be visible to a machine, that the future of how we will feel is naked before the soulless red eye of a bestseller bot.

Consider instead that the secrets of great art have in fact *always* been out in the open, coming into and out of shots too quickly for us to make a mental record of. We now have a tool to help us catch those fleeting movements and make them into more beautiful and moving works. Novelists and screenwriters understand that the most trying aspect of the creative process is decision making. Each choice affects every other one in ways large and small. A tiny error can force a concatenation of bad decisions. Any tool that reveals to the writer the cost of a particular decision is a useful one.

The idea that art is beyond judgment, that it comes from a unique place within the human soul, comes from the Romantic period. It's a useful ethos for anyone actually trying to make art. Faced with the task of pulling beauty and truth from the ether, the canvas must be either a vast plane of possibility or else it will be prison. But ultimately, outside all the frustration and egotism that is involved in the act of its creation, art is a dialogue. Like any exchange of ideas and emotions, the effects of this conversation can be measured, and thus predicted.

A screenplay, too, is a machine that is part of a larger apparatus. It's no wonder we often refer to characters, tropes, and plot developments as devices: they are precisely that. Once the screenwriter has undertaken a wide number of decisions, several readers must then decide whether to pass the script along up the ladder (not without some personal cost) or to kill it. At the pinnacle of the long, strange dance known as the green-lighting process, a studio makes a grand choice about a project and all the other decisions that project entails. A new machine is set in motion, that of production and release. Once it's concluded, the budget has been allocated and spent, the actors have made their way across the set, the helicopters demolished, the gratuitous nudity edited out (or in), the marketers have done their marketing, the result is presented to the moviegoing public at movie theaters around the world.

And there stands Jehoshua Eliashberg. What he offers is nothing more than a view, a set of inputs to change the decision-making process, a feedback mechanism. Hollywood, video-streaming services, and moviegoers are presented with a new choice: whether or not to listen.

CHAPTER 6
The Spirit of the New

THE lobby of the Saatchi & Saatchi building in downtown Manhattan prominently features three enormous multimedia paintings by Frank Stella. They explode off the marble wall in a shriek of color and movement. These were procured in the 1980s during the financial peak of the New York art scene. The '80s also marked the apogee of American advertising, and the building's namesake, the Saatchi & Saatchi firm, occupies the top floors.

The view from the company's office is clear over the Hudson River. The waiting area is sleek and modern. When I arrive I discover a handful of young, attractive executives in a meeting room. They're judging radio spots. The men wear V-necks beneath their tweed sport coats. Their tapered, skinny jeans descend toward canvas shoes. Everything about this scene, and the people in it, suggests a bright present branching toward a better future. Probably no organization in the world understands the value of appearances better than Saatchi & Saatchi, and so it's not surprising that the first thing a visitor to their offices will encounter is a subtle advertisement for the firm itself.

Make no mistake, the company's wrinkles have been Photoshopped away and Saatchi & Saatchi is running on borrowed time.

I meet Becky Wang, the company's director of insights and analytics. Wang uses market research; Facebook and Twitter posts; cookie, click, and app data to help Saatchi create better marketing campaigns. Some of these ad campaigns still come in the form of one-way ads to generic customers during football games, in magazines, on the radio. These are just a bit smarter than they were ten years ago. Other messages don't fit with the classic definition of a commercial at all. They're Web and iPhone apps, tailored promotions, and digital interactions that collect information as much as they pitch specific products.

Becky Wang, her clients, and the other smart folks trying to sell things across the digital landscape know that the latter—apps and interactive sites that cost little to create—are becoming more important to clients than TV spots that can run into the millions.

"The advertising industry was starting to move behavior, but we used to think that if you told someone something and evoked some sort of emotion, that was enough. We now know that's not the case," she says.

The premise of traditional advertising is that exposing consumers to your product or brand repeatedly through pictures and images will get them to buy it. This is a flawed perception. A 2010 McKinsey study of twenty thousand people showed that the process by which consumers discover and develop an appetite for new goods is now circumventing such traditional outlets as radio, television, and magazines.[1] Today, consumers hear about new products through targeted online ads, searches, and friend recommendations.[2]

The study also found consumers walk around with a broad familiarity of all sorts of brands all the time. We know what Coke, Samsung, Apple, and American Airlines are and what they do. Therefore little commercial reminders about the existence of American Airlines can be a nuisance. It's only when consumers enter into a state where they're actively looking for a product, when they're "in the market," that advertising actually brings the consumer much closer to pulling out her wallet. This underlines a fundamental dysfunction in traditional advertising. A Super Bowl spot can't measure whether the audience is actively looking for the product being pitched. Chief executive officers are growing increasingly fed up with throwing money at marketing and ad firms. Two-thirds of the CEOs polled by the Fournaise Marketing Group said that they doubted the efficacy of traditional marketing and felt there was no clear connection between ad spending and sales.[3]

CEOs showed that they felt like they were investing in ad campaigns but there was no way to track with precision how those campaigns were bolstering the bottom line. That may not sound like a new predicament for marketers and advertisers but today CEOs who want to break up with Madison Avenue have alternative places to invest their money. The idea that big data poses a threat to traditional marketing practices has been so widely accepted that software firms such as Adobe use it to actively market their big data solutions to marketers.

In his 2011 book *The Daily You,* communications expert Joseph Turow documents exactly how the decline of traditional advertising began almost unnoticeably in the 1990s when more and more content providers such as newspapers and magazines began going online where engagement could be tracked through click-throughs, rather than speculated on via circulation or Nielsen ratings. But the single biggest factor in the death of the *Mad Men* model was the rise of the cookie, a little piece of software code that embeds itself on your computer and records your movements around the Web. Thanks to the rise of such integrated ad networks as Yahoo!, personalized ad delivery platforms such as Google's AdSense program, and data-exchange companies, ads now follow you from site to site and collect information about you as you go. It's the cookie that enables Google, Yahoo!, and Facebook to show you ads for vacation homes in Florida because you once looked at the South Beach, Florida, page on TripAdvisor.

Turow calls this the "decoupling of audiences from context."[4] One effect of this decoupling is that ads that transmit a carefully conceived and constructed image in one direction, the sorts of images and campaigns that built the Saatchi & Saatchi fortune, have become financially less attractive no matter how pretty those pictures are to look at. When Saatchi receives less money for photos of Jennifer Aniston holding Louis Vuitton handbags, magazines charge less money for ad space, both in print and online. That means they have less money in general for reporters, fact-checking, and making product.

Today, magazines such as *Wired,* the *Atlantic,* and my magazine, the *Futurist*, deliver important stories and beautiful photos in an online environment even faster and better than we did in print. Unfortunately, it doesn't matter; as an editor, you can invest the last of your budget in an eight-page piece on aquifer depletion but when the reader moves away from your Web site to go look at Tumblr blogs of cats in Halloween costumes, the ads follow her. Why should a company buy an ad at one location when it can buy access to a person's attention virtually anywhere?

The migration of content from print to the Web is not a good thing for anyone who cares about great journalists doing good work. But the shift is a positive one if you're a company interested in acquiring lots of behavioral data about where online your present and future customers go.

Advertising is just one of several components that drive product sales. The others include price, placement, demand, and the quality of the product itself. All the gears have to be working in concert for a sale to happen. But the amount of information about how those components interact is exploding, which is changing how clients value creative firms like Saatchi & Saatchi.

"We're no longer held to one basic metric like how sales or market share improve" as a result of an ad campaign, says Wang. It's a change she welcomes. Advertisers, she says, have long felt they get the blame for things they can't control, how products are designed, placed, and priced. Today, "We're held to how much time customers spend on site or with brand. If I have three hundred thousand moms actively using an app that becomes a success metric, I'm no longer beholden to something I have no control over."

One example of a product that reveals the importance of *time with brand*, as opposed to *sales*, is rap artist Jay-Z's album *Magna Carta . . . Holy Grail*. Electronics giant Samsung partnered with the artist to promote the release, scheduled for July 4, 2013. Samsung, which makes the Galaxy S4 smartphone, announced they were going to give away downloads of the album to the first million individuals who downloaded a free app for the phone. At first glance, this would seem to be an extremely generous move on the part of Samsung. The music industry may be in a shambles but surely Jay-Z can still sell a million downloads on the opening week of a new album, so Samsung was leaving money on the table, right? Sometime between downloads one and five hundred thousand, Atlanta rapper Killer Mike took a look at the terms of service agreement for Android users, which stated that the Jay-Z Magna Carta app needed access to the phone's system tools, network communication records, phone calls, GPS location, and more. This prompted *Gawker* writer Adrian Chen to ask, "Why does Jay-Z need your GPS location? Is he going to cruise by on a platinum-coated jet ski, personally chucking out copies of the album to people who downloaded the app?"[5]

Jay-Z naturally has no interest in your location; marketers working with Samsung do. Another example of how continuous user data is becoming more valuable than a one-time product sale is Nike+, a self-tracking system much like Fitbit. It allows you to create individual data streams of your activities. You learn about how you run, what conditions work best for you, and you can compare your scores with those of your friends or people around the world. If you go in for all the available Nike+ merchandise it becomes an expensive system. There are wristbands, clips, and the cost of constant upgrades. Not surprisingly, Nike wants people to interact with the system as much as possible, so the company created a set of little games in which people could take part. They can compete against their own scores or against other people with Nike+ watching to see how, when, and how often people use the Nike+ system.

Nike learned that people who didn't take challenges didn't continue to use the shoes. Suddenly, selling the Nike+ system wasn't enough anymore. The company also had to sell consumers on using it the right way, in the way that gave Nike actionable data and encouraged customers to use the product continuously, to make a new habit of it.

For an advertising company, pitching the right way to use a product such as Nike+ isn't necessarily easier than pitching the product itself. People don't adopt new habits simply to suit sports apparel manufacturers. Habits, even the ones we don't like, are personal. "We are what we repeatedly do," said Aristotle. But information about what we do is now part of our digital trail and thus our naked future. For advertisers, understanding those often repeated behaviors is the first step toward changing them.

In March 2011, before Wang joined Saatchi & Saatchi, she collaborated with Drew Breunig, a young technology director at a data analytics group called Annalect, on another project. The client in that case was an antismoking group looking to discover what sorts of conversations people were having on Twitter about tobacco, and whether those conversations could be used to predict something specific about smoking behavior.

"We created an entire dictionary of smoking-desire statements, like 'Man, I really want a cigarette right now,' or smoking-consumption statements like 'I'm having a cigarette outside,'" says Breunig. "We counted all the synonyms, all the different statements that would fill up this dictionary." They wound up tracking consumption of a whole host of compulsive substance habits, not just smoking but also alcohol and caffeine intake. When Breunig and Wang applied the dictionary to a large data set of tweets from New York City, they found that Monday through Tuesday on weekdays, coffee drinking would peak and go down right around two o'clock in the afternoon. Alcohol consumption would start around seven o'clock, peak at about ten, and fall off a cliff at two in the morning.

"Cigarettes were the liminal vice," says Breunig. They came between consumption of coffee and alcohol. When it was too late in the day to have a morning cup of joe, people tweeted that they were grabbing a smoke and then, later, that they really needed to try and quit smoking . . . while they were smoking, outside a bar between one and two in the morning.

For an advertiser looking to place a digital ad across an ad network at a particular time, that's extremely useful information. "If I can predict you're a smoker, I can predict that you're going to have a craving during weekdays, right when it's too late for coffee. So if I'm the [antismoking group], and I want to help people not be smokers, I'm going to buy every banner ad I can in that time slot and these ads will say, 'Hang in there, man,'" says Breunig. It's a good thing Drew Breunig's clients wanted smokers to quit.

Surprisingly, data-driven marketing didn't begin in the Internet era but long before. To understand how we got to this particular point, where communications scholars are writing eulogies for advertising firms and start-ups are predicting when you will want a cigarette, you have to go back several decades.

The future of advertising resembles its past.

What Happens in Vegas Will Follow You Everywhere

In the mid-1990s Gary William Loveman found work at Harrah's casino chain where, in just a few years, he changed the industry forever. With a PhD from MIT's Sloan School of Management and a résumé that included work at the Federal Reserve Bank of Boston, he could have gone anywhere. But at Harrah's (now owned by Caesars) a brilliant quantitative mind saw a field full of low-hanging fruit. Loveman was not a creature of Las Vegas but an analytical anti-gambler, a man with a deep suspicion of hunches. The customer-service-driven casino environment offered the opportunity to apply his quantification mind-set to an area dominated by big personalities, intuition, and instinct. "There was very little formalization of service as a *discipline*," Loveman told Karl Taro Greenfeld of Bloomberg News in 2010 (emphasis added).[6] "There is a long history of research and engineering around factory optimization, scheduling, and throughput. On the other hand, the service sector was seen as a poor cousin."

The Las Vegas casino business model before Gary Loveman was built entirely around the concept of bigger: bigger signs, bigger fountains, bigger volcanoes and attractions visible from the Strip, bigger names on the marquee to pull people in, bigger lobbies, more of everything. In the fight for flash Loveman saw a war not worth winning. He turned his focus to the job of remaking Harrah's utterly unglamorous customer loyalty program, transforming it from a simple thanks-for-coming voucher scheme into a massive, data-run, telemetric decision engine.

Loveman didn't invent the customer loyalty program; the airlines did. In 1978 the U.S. airline industry was deregulated, big carriers expanded their routes and slashed their prices, and a vast frontier was suddenly open. American Airlines relocated to the Dallas–Fort Worth area the following year, where an executive quickly realized that the company could offer lower-price fares for the customers that used the airline most often. But finding these people was no easy task. Fortunately for American Airlines, they had one of the world's biggest computerized databases, the Semi-Automated Business Research Environment (SABRE). SABRE allowed travel agents and American Airlines' clerks dispersed around the world to book passengers on quickly filling flights in something like real time. In what might be considered the first case of a major company using a computerized database for customer profiling (outside the insurance industry), American Airlines scanned their database to figure out who were their 150,000 best customers.[7,8] These people became members of the world's first computerized customer loyalty program, AAdvantage, in 1981. (Frank Lorenzo, Texas International Airlines CEO, came up with the first frequent-flyer club in 1979, but this small airline lacked the computer resources of its larger competitors.) After the official launch, other big airlines developed their own programs within a matter of days. These soon included not just deals on airfare but also rental cars and hotels in such trendy spots as Las Vegas, Nevada.

Loveman took what the airlines had been doing for decades, and was in place already at Harrah's, and perfected it. In 1998 he created the Harrah's Total Gold program, today called Total Rewards. Here's how it works: when club members book their hotel or restaurant reservations, when they swipe their Total Rewards club card in one of Harrah's video slot machines or use their Harrah's account to place table bets, when they win, when they lose, when they hesitate, when they cash out, feel the itch, and come back to hit the one-armed bandit one last time on their way out of town, the system knows . . . and remembers. But the database is more than just a play-by-play record of plundered 401Ks. Customer service reps both in the casinos and at call centers around the world look to learn everything they can about Harrah's Total Rewards members to tailor very specific offers to them.

It's not cheap to collect, keep, and utilize all this data. Harrah's reportedly spends more than $100 million a year on IT, but Total Rewards has more than earned its investment back. What started with 12 million subscribers in 1990 hit 26 million in 2003, more people than the combined populations of Greece and Portugal. By 2010, 40 million people were in Harrah's system. This gives the company access to the lives of the people in its casinos. Instead of rigging the table games, the system works to rig the customers.[9]

"Let's say we have a sixty-year-old woman who lives in a comfortable suburb of Memphis," Loveman told journalist Robert Shook in 2003. "She visits our Tunica property on a Friday night, briefly plays a dollar slot machine, and goes home. Based on traditional casino methods, she'd have a low theoretical worth, perhaps a few dollars. Consequently, she wouldn't be a likely prospect to pursue, and little effort would be made to get her to come back. And in all likelihood, she wouldn't respond to it. Our present system draws distinctions between the observed worth of a customer and what we predict their worth is."[10]

Your predicted worth is a number representing how much money the Harrah's system has calculated you can be persuaded to lose—er, gamble with—when you go to one of their locations. It's based on your ZIP code, what you play, how you play, and other indicators of wealth and willingness to gamble. Reportedly, the system is 90 percent accurate at predicting how much a customer can be persuaded to drop at a casino.[11]

These customer-worth scores serve a greater function, predicting exactly what offer potential customers respond to and when to issue it. If you're in the casino and you're losing heavily (as measured by the account activity on your card), Harrah's will dispense a "luck ambassador" with a coupon for something the system has calculated you'll like, anything from a free drink to show tickets. But these offers don't just come to you while you're on the floor. If you go to Harrah's on your birthday, on your annual vacation, on the seventh day of the seventh month of the year, the company will send you an offer timed to encourage you to do that sort of thing again.

Do you go to the casino when your wife is out of town? As Christina Binkley of the *Wall Street Journal* discovered, Harrah's knows that, too. She observed a Harrah's-hired telemarketer named Mr. Salvador go through the process of contacting repeat customers with special offers and saw firsthand the wealth of data Harrah's can use to turn any customer encounter to the company's favor. "On a recent list was a thirty-four-year-old man who hadn't been to a Harrah's Entertainment Inc. casino since November 2003. Before then, according to the data, he had made trips to the Rio in Las Vegas, as well as casinos in Tunica, Miss., and East Chicago. 'This is a customer who can only play when his wife is on vacation or when he's on a trip,' says Mr. Salvador.' "[12]

Where did that information come from? There are several ways to infer it. This thirty-four-year old man might have indicated on a customer satisfaction survey form his marital status and that he was traveling alone. Alternatively, perhaps when he entered that Harrah's hotel in November 2003 a friendly desk clerk or casino cashier asked if he was in town on a special occasion and he mentioned that the special occasion was being away from his wife. She simply complied with the company guidelines and typed this answer into the form in front of her. Either way, what was a very forgettable exchange for this thirty-four-year-old man has become a piece of data that now follows him everywhere and influences how and when Harrah's contacts him. Harrah's has become a vast sense organ. As an institution, it's constantly detecting and responding to new information on its customers. It remembers everything, weighs every interaction. It knows your limits better than you know them yourself, and it wants you to keep playing.

Writers in the business press credit Loveman with completely changing the culture at Harrah's. Read that to mean he fired a lot of people, mostly in marketing. In his interview with Greenfeld, Loveman explains his impatience with traditional marketing methods. "Testing and measuring is important to us. When our employees use the words 'I think,' the hair stands up on the back of my neck. We have the capacity to know rather than guess at something."

What does the history of the casino loyalty programs mean for the future of shopping, marketing, and advertising? Simply put, the retail world of 2013 has become a Harrah's casino.

Futurizing Tuna Steaks

The date is April 25, 2012. I'm at an upscale gourmet grocery chain in a tony Toronto neighborhood. With me is Wojciech Gryc, the twenty-five-year-old creator of a software platform that automates customer analytics. His company, Canopy Labs, is looking to put the predictive analytics-crunching capabilities of Walmart and Target into the hands of small- and medium-size businesses. The Canopy Labs platform "predicts which customers are most likely to accept the offer you are going to be suggesting," he explains. Though only four months old, Gryc's company has already attracted funding from Silicon Valley angel investors. He's riding the big data hype wave but his tech experience is deeply ingrained. He got his start in big data at IBM but his dad had him at the computer at the age of ten. Point to a product and he can tell you how to use data to sell it. "You can optimize any shopping experience," he says. In this instance he means optimizing the experience for the seller, the clients he works for.

We're here to see what that looks like.

Grocery stores are laid out according to a marketing science that's actually been around for decades. Thousands of recorded customer decisions help such outfits as the middle-class gourmet grocery that I'm standing in to nudge customers where the store wants them to go. Past the cash registers we encounter cakes and cupcakes. These are prominently displayed because they're potential impulse buys. They have a higher profit margin and longer shelf life compared to the fruits and veggies. They're also the sort of thing people don't put on a regular shopping list. The store wants to make sure we see them. This is an act of priming. Even if you demure the sweets, you still had to think about it. You'll be less able to fight off the next impulse.

Over the past decade no chain amassed more data to figure out these sorts of placement decisions than Walmart, which experiences more than 1 million customer transactions every hour.[13] In 2004, in what is probably the most famous instance of a retailer using customer data to effect product placement, Walmart stacked the shelves of its stores standing in the path of Hurricane Francis with Pop-Tarts and beer. Their terabytes of customer data from a previous mega storm, Hurricane Charley, showed that these were the items most sought after in Walmart stores before hurricanes. Hurricane survivors, it seems, like treats that can be microwaved or ingested straight out of the package (only the latter is advisable in the case of beer).

Discovering the reasons why customers pick up one item and discard another is the central challenge of predictive analytics in retail. This is what I call the Bear problem. My cat, Bear, rejects food randomly. I open a can of a type of food he liked yesterday, or a week ago, and he'll either turn his nose up at it or eat it for reasons that aren't obvious. He also spends part of the day running around outside. We have an inside cat who never rejects food. If we assume that Bear's fickleness is caused by something he discovers outside (which is indeed just an assumption of causation based on an observed correlation), then determining why my cat rejects what I give him involves tracking a huge number of variables, such as where he goes on his excursions and what he encounters there.

In figuring out why people pick up one thing and put down another, Walmart is basically faced with the same problem that I am. And they've employed some controversial tricks to get at it. In 2003 Walmart and Procter & Gamble ignited a firestorm of controversy when the *Chicago Sun-Times* revealed that the two companies were outfitting cosmetics with RFID tags. When customers picked up one of the chipped cosmetics items like Lipfinity lipstick, a surveillance system would follow the customer around the store to see what other items she considered buying but did not.[14]

Both Walmart and P&G maintain that they used the tags solely for the purpose of tracking how the consumers handled the products *inside the store* and, it should be said, no one has found any evidence to the contrary. But from a retail perspective, data about how customers use products outside of a store are far more useful than knowing what they do inside the store. Does the average Lipfinity owner leave her lipstick on her dresser at home or carry it in her purse? Where does that purse go?

Not of all Walmart's tactics to triangulate customer behavior have been so controversial. In the fall of 2006 Walmart stepped away from the stereotype of the impersonal big-box retailer with the creation of its "store of the community" program, which gave managers much more leeway in terms of stocking their shelves and laying out their stores. The company took careful note of what worked in what market, to better understand why some strategies succeeded and others failed. This enabled Walmart to expand the number of planograms, or acceptable merchandise-layout configurations, from five to more than two hundred by 2010, an important step in customizing the retail experience to the individual. What was an impersonal big-box outlet began to metamorphose into a village store, better reflecting the purchasing habits of the community.[15]

But altering the layout of particular outlets and the displays therein for *every* shopper will never be practical, even in the future. The challenge today is to re-optimize an experience designed for a statistically generic person into an experience for living individuals.

How do you personalize something that was designed for the aggregate?

"If you were the management of this store, how would you optimize tuna steaks to me?" I ask Gryc. We determine that one of the key variables of tuna steak pricing should be freshness. Tuna steaks are highly perishable. From the store's perspective, it makes better economic sense to offer me a discount on the tuna if the alternative is throwing it out at the end of the day. In the vast majority of grocery stores around the world, the current method for accomplishing this is to hang a SALE sign on the tuna's glass casing. A better method, Gryc explains, is notifying your customers who buy a lot of tuna and then offering them a time-sensitive coupon. These e-coupons can easily be delivered to a user's phone on the basis of context, meaning where that user is and even what activity that user is engaged in, easily determined by other app usage or simply location. (Someone in a bowling alley is unlikely to be windsurfing.)

The store can then track the number of those coupons that are redeemed. This is sometimes called one-to-one marketing at scale. It allows the store to order tuna with greater confidence that it'll be able to move it at different price points. It's less likely to have excess tuna and won't need to offer as many deep discounts. A restocking decision that, perhaps, was once made at the district manager level can be made at the department manager level; instead of being made in a distant office, it's made by the guy behind the counter. When people redeem the coupons the store gets data on which sorts of customers respond to which sorts of offers.

Personalized, in-store coupons have existed in various forms for years. In 2006 the Stop & Shop chain outfitted the carts of three of its Massachusetts branches with a digital personal assistant called Shopping Buddy. You just swiped your card and received personalized discounts and offers while strolling the aisles. It was just like shopping at Harrah's![16]

The drawback, of course, was that you had to be in the store to get the coupon; the program didn't work to attract the sorts of people most likely to accept offers.

Today, people carry their own shopping buddies in their pockets. The smartphone has become the essential shopping accessory. In 2012 more than a quarter of smartphone owners used their phones in stores to read reviews and to hunt for better prices on the goods.[17] Store-sponsored smartphone apps respond to this trend by offering personalized coupons to customers where they are. Today, all sorts of stores offer a variety of different app-based personal shopping assistant programs for iPhone and Android, which interact with customer loyalty accounts.[18]

UK-based retailer Tesco can track 80 percent of its sales through its club card and could provide more than ten coupon variations in 2012. They, too, tailor deals to individuals on the basis of inferred net worth. Even Walmart, with its business model of always having lower prices than its competitors, found enough wiggle room in its pricing structure to offer extra-special low prices to folks willing to give up a bit more in personal data. A couple of years ago, Walmart put together a customer loyalty program called eValues, which targeted specific deals to specific customers through e-mail and apps.[19]

"It's kind of like the eHarmony of couponing—we find the very best offers for the customer," Catherine Corley, vice president of member program development at Walmart, told the *Daily Herald*.[20]

These programs, and the individualized offers associated with them, would seem to be a victory for consumers. That quality of *seeming* is important. People who participate in customer loyalty programs actually spend more at stores they shop at than people who aren't part of such programs—the same way people in Harrah's Total Rewards wind up gambling, and losing, more at casinos than those who come to the casino with only cash.

Before long, eValues customers were making twice as many trips to Walmart as people who weren't in the program. Walmart was willing to slash its low prices even further for the same reason

Harrah's likes giving away hotel rooms to little old ladies. Both were looking at the long game, what your consumer behavior looks like over time so they can predict what sort of customer you will be in the decades ahead. Today, eValues is called Instant Savings and it's available to Sam's Club members (Sam's Club is owned by Walmart). Various other aspects of the eValues program have been rolled into the Walmart and the Sam's Club apps.

Naturally, your customer data belongs to you first. You are the point of origin. And with just a little effort you can get a sense of how the stores that you shop at, such as Walmart, view you and your lifetime value as a customer. If you're interested in performing this search on yourself, you can go to the investor relations portion of a company Web site, request an investor prospectus, and find a profile of an average customer to see how you compare. Publicly traded companies have to release annual sales figures, profits, and liabilities and these often include information on target demographics. You can also go to the Securities and Exchange Commission's EDGAR database and search for a particular company's 10-K form.

For instance, the average Walmart customer spends $1,088 per year at the store, makes twenty-seven shopping trips in that year, and spends $40.30 per trip, according to the most recent publicly available information.[21] If you spend more than that at Walmart, you have some idea how important you are relative to the average.

Do you know if you're part of a demographic that the store is going to court more aggressively? That can be a factor as well. Walmart (publicly) divides its shoppers into three groups: "brand aspirationals," people without much money who don't want to look cheap and so buy brand-name items at discount prices to cover that up; "price-sensitive affluents," meaning cheap rich people; and "value-price shoppers," regular cheap people.[22]

Your ZIP code could also be a factor in how you're scored. Big companies use geo-information services (GIS) to figure out the income levels for different neighborhoods. One company that provides both GIS software and GIS insights is the Environmental Systems

Resources Institute (Esri). It can classify any particular neighborhood into sixty-five different segments on the basis of income, consumer habits, number of kids, average level of education, as well as dozens of other variables, and does this on a block-by-block basis. (The information comes from the U.S. Census.) Within these segments is a fair amount of nuance. People who fall into the "military proximity" group are twenty-eight years old on average, make $41,000 a year, don't have pets but do have renter's insurance, and go to places like SeaWorld on vacation. "Great expectations" are people who make $35,000, live primarily in the Midwest, and do most of their grocery shopping at Walmart. For big businesses looking to enhance customer targeting, this is immensely valuable information. Esri makes a lot of this data available to anyone through their premium Web product, the ArcGIS platform. It's not free but Esri does offer free trials and extremely generous pricing for nonprofits. Keep in mind that your neighborhood score will be used differently depending on the business you're looking to engage with. For instance, if you want to lower your insurance premiums, don't buy a house in a "good" neighborhood if it's also an expensive neighborhood. Instead, move next door to clerical workers.

As mentioned earlier, if you use Verizon or AT&T, your phone company is also helping marketers much better target mobile ads to you. AT&T, for instance, offers a product called AdWorks that promises to "connect advertisers with their audiences across online, mobile and TV channels." In other words, it helps advertisers stick particular ads in front of your face depending on where you are and what you're doing.

To do that, AT&T partners with data brokerage companies such as Acxiom. You've probably never heard of Acxiom but rest assured, the company has heard of you. Acxiom has information on more than 500 million people around the world, an average of 1,500 data points per individual, around 6 billion total pieces of information across all of Acxiom's databases. This data could be anything from the magazines you subscribe to, to the sort of car you drive. It's information you volunteered on surveys and when you opted in to various service contracts but much of it was just sitting in the public domain. Acxiom uses that to put you into one of seventy different customer classes based on income, education, and other factors. That's important, because Acxiom, AT&T, and Verizon can't sell advertisers access to you *specifically*; that would be a clear violation of privacy. They sell you as part of a group of people sharing certain characteristics. And no matter what group you are in, they are extremely skilled at finding you. Acxiom knows how many people in every one of its clusters are reachable via mobile phone, browser ad, or television ad *at any given moment.*[23] Let's say you don't click an ad when it shows up on your phone or on the Web but you still want the product. You go into the store and buy the item there. Acxiom knows that as well. The ads you see don't just follow you as you go from site to site, they follow you everywhere. But Acxiom isn't just selling advertisers access to the people in your cluster, they're also selling your future decisions.

In March 2013 the company released a new product: Audience Propensities. A propensity is a prediction, hedged by a probability score, about a specific consumer behavior, such as how a customer will respond to a particular offer. For instance, let's say you have a discounted insurance product and you want to reach only those potential customers who would be extremely unlikely to buy that product at full price. Acxiom executive vice president Phil Mui and his team showed how Acxiom identifies the people with this propensity in a live demo at a product briefing in March 2013. In a manner of minutes the Acxiom system crunched 700 million rows of data and outputted a number. Mui revealed to the audience that if they were looking for someone with that propensity, "there are 275,012 people that you can reach out to." Mui was careful to point out, "This is live. You can buy this audience *today*." There are three thousand such propensities Acxiom can model.[24]

In September 2013, after a spate of unfavorable press and inquiries from the U.S. House of Representatives, Acxiom took a bold move in the right direction and opened a Web site called Aboutthe

Data.com, to give consumers a partial window into the sort of data the company had on them in its databases, afford consumers the power to make amendments to their profile (these are tracked), and even opt out of being in the Acxiom database. The move was not without risk for Acxiom. As company CEO Scott Howe told *New York Times* reporter Natasha Singer, "What happens if 20 percent of the American population decides to opt out? It would be devastating for our business."[25] The database is a good first stop for anyone looking to better understand how she looks to marketers.

Careful record keeping and a bit of calculation can give you an idea of what companies often refer to as your "lifetime customer value." But this is only an idea. Past behavior doesn't dictate future results. You may have a cat today, but what if you fall in love with a dog person, or a ferret person? (Don't do that.)

The big data present can give retailers a good understanding of your future buying as your future exists right now, but the naked future is one that's always moving. To gain an understanding of what that movement looks like, you need more than a snapshot; you need to understand how the subject you're observing is evolving, where she goes, what she does, what she encounters. You need to know who she talks to, who can influence her, and whom she can influence.

The days of planting RFID tags in cosmetics are long gone. Such cheap tricks are no longer necessary. Today, consumers give that information away eagerly.

How Facebook Turned You into an Advertisement

If you were on Facebook sometime between August 14 and October 4, 2010, you probably played a role in an experiment. Facebook turned 253 million users into test subjects to study contagion. No, Mark Zuckerberg didn't release a hostile virus into the New York water supply (yet). The contamination event that the Facebook Data Science Team was monitoring was related to information, specifically URLs and how they spread between people.[26]

Here's how this experiment worked: You were randomly placed into one of two groups. If you were in the first group, then when one of your friends posted a story you saw it in your News Feed as you normally would. If you were in the second group, the same story would appear far lower down in your News Feed where it was much less likely to be seen.

The objective of the experiment was to examine the probability of a user's sharing a news article, video, or link even if that user didn't know anyone else who had shared it. Facebook's interest in information contagiousness goes beyond curiosity. A customer's friends' Facebook posts are an indicator of—among other things—how likely a customer is to abandon a company or brand or pick up a new habit.

The head of the experiment for Facebook was Eytan Bakshy, one of the star players on the Data Science Team. In person Bakshy seems very young to have such a coveted job, with access to a user base of hundreds of millions of people to experiment on. He bears a strong resemblance to the character of Leonard Hofstadter, the experimental physicist character played by Johnny Galecki on the geek-beloved television show *The Big Bang Theory*, but Bakshy comes off as a bit more serious . . . and a bit smarter.

The Facebook team knew that if they could show the likelihood of a user's sharing an item when none of her friends had shared it, then the team could show how much more likely a user is to share a link that comes to her from someone in her network. And getting people to share information within a growing network is the entire value of Facebook. The ability to prove that the Facebook News Feed and the information shared in it can cause a behavior change is an extremely important aspect of the Facebook business model. Bakshy and his team found that you're 7.37 times more likely to share a link that one of your friends has shared than to share that same article with no social signal.

The experiment also gave Facebook insight into a far more difficult question, one with a more potentially lucrative answer: how your relationship with different people influences the likelihood that you will like what they like.

When it comes to purchasing behavior, understanding who is influencing whom is a murky question because of a phenomenon called homophily, which is the tendency of similar people to exhibit similar behavior. If you and I both attended a liberal arts college, are of the same income, work in similar professions, and share some other overlapping demographic characteristics, there's a good chance that we'll both post a big article that appears in the *New York Times* to our Facebook page independently of each other. If the article shows up in my News Feed before it shows up in yours, it's not clear that I influenced you to post it. You might have just happened to see the same article that I did later in the day. The Data Science Team's experiment provided a formula for determining who in a network is inspiring who to share what. But the study's most surprising revelation was that the people you're closest to don't necessarily influence your online behavior more than the people you're only nominally friends with, folks with whom you have only a casual off-line relationship if any at all.

Facebook, it turns out, is particularly useful for researchers looking to separate fake friends from real ones. In sociology these two groups sometimes go by names that were originated in 1973 by sociologist Mark Granovetter: weak ties and strong ties. Our strong ties are the people with whom we interact often. These are family, friends, people we have a lot in common with, and with whom there is much homophily. Our weak ties are the people we add to our network without quite knowing why. Off-line, that category includes people you've had sparse interactions with, the girl from that cocktail party whom you remember as interesting, even if you can't remember her name. Before social networks these weak ties would orbit for a while and then be lost to oblivion. On Facebook your weak ties remain visible through the News Feed even if you interact with them very rarely.

The experiment showed that weak ties in *sum* exercise more

effect on your sharing than even your friends and family. They serve an important role introducing us to news and stimuli outside our normal circle. "These people share information from sources we might not frequent. As a result of seeing content from these users, you are many times more likely to share it. Surprisingly, while strong ties are individually more influential, weak ties are collectively more influential," Bakshy told the crowd. Despite the reputation of Facebook as a place where people post mostly personal pictures of their kids and cats, in fact most of the links, articles, and other content that people share come from weak ties. It's not personal, but news, petitions, meme photos of a grouchy cat, stuff from outside. But it has allowed Facebook to better anticipate what sort of content, what memes, provocative blog posts, and other material its individual users will respond to, Bakshy says. The evidence suggests that the approach is working. For all the recent talk of Facebook being the passé social network, it continues to dominate in a number of key metrics, perhaps the most important of which is time on the Web site. The average Facebook user spends almost an hour a day on Facebook, far more than users spend on any other social network. No, our relationship with Facebook is not as exciting as it was. But it is stable, even marital.

It's also allowed Facebook to better predict which of the people in your network, whether real friends or weak ties, can move you closer to buying something, and by how much.

Skip ahead to a second experiment that Bakshy spearheaded in 2011. This one involved a far more modest subject pool of just 23 million Facebook users. It worked like this: the subjects saw a story in their News Feed, perhaps for a History Channel broadcast or the Tough Mudder decathlon sporting event. Unlike a provocative *New York Times* piece, a funny George Takei photo, or some bit of information that you may share naturally, the purpose of this story was clearly mercantile. It probably didn't look like a commercial but it was still a product endorsement, or, as Facebook calls it, a sponsored story.

The subjects of the experiment were placed into three groups.

The first group saw the story and the identity of one friend who liked the associated product. The second two groups saw the story and the identities of more friends who liked the product. The stated goal was to measure the role of "social influence in social advertising."

In advertising, well-paid celebrities have long been a proxy for the familiar. Psychologists Carl Senior and Baldeesh Gakhal have shown that we're more likely to buy something from a famous person than from someone who is merely beautiful, in part because we trust familiarity over physical attractiveness. No one is more familiar to you than your friends. That's why they're better pitchmen than virtually any celebrity.[27]

What Bakshy and his team found is that even the *slightest hint* of weak-tie friend affiliation with a product or brand can increase the probability that a Facebook user will "like" a product story by 10 percent. The effect is much more robust when users are informed by Facebook that one of their *strong ties* likes something. Now here's where contagion comes in: once you click "Like," a seemingly innocuous action that the entire Facebook platform and *all* of its affiliates prompt you to do all the time, you become an advertisement to your friends. They see the sponsored story about the product you "endorsed." A number of them see the link and also like it; next thing you know everyone is taking Nike+ challenges and has no idea why.

Bakshy and his team can also measure the dose-response function of each sponsored story to which you're exposed, in effect predicting your tolerance rate for this sort of marketing or even, one day in the not so distant future, how quickly certain friends exhaust their influence on you. Bakshy has indicated that this is a possible future direction for the research. After all, there are thousands of different ways people can be tied to one another and, within the context of a social network where every interaction can be seen and scored, many different ways to measure those interactions—including, in the words of Bakshy, "trust and intimacy."

Facebook has already begun making use of these insights. A program called Facebook Offers lets businesses extend coupons and deals directly to fans through the News Feed. When Facebook first announced offers, users couldn't control whether their friends saw if they accepted the offer; your acceptance was a tacit endorsement that was broadcast to your friends.[28] In the summer of 2013, Facebook second-guessed this approach. Turns out when users agree to share the fact that they accepted an offer, they have a much bigger influence on their friends to accept that same offer. [29]

If you find yourself in a strange city, a program called Facebook Local Search, currently part of the company's mobile app, can give you a list of local places to visit. Today, it works a bit like a less fun Foursquare. But Facebook collects a lot more personal and connection data than its competitors. In the future Facebook Local Search and Facebook Home suggestions will be based on your habits, your likes and dislikes, and which friend recommendations are going to be most influential for you. You, too, will be making recommendations, perhaps without realizing it.[30]

Facebook assumes that you'll come to appreciate the additional personally relevant context in the ads to which you're exposed, the additional convenience of knowing that someone in your network liked a product that you're considering. They may be right. We may not yet feel at ease with the direction that advertising is taking but few of us have a strong attachment to its current manifestation in which we're constantly bombarded by images and sounds from people we do not know, selling us items we do not want, and incapable of hearing us when we voice our refusal.

The big-box retailers are trying to shrink themselves down to like size. In 2011 Walmart purchased Silicon Valley data-mining company Kosmix to help Walmart Labs get a handle on social network data from Facebook and Twitter. They deepened their investment in 2013 with the acquisition of predictive analytics start-up Inkiru. They're competing with dozens of other outlets and corporations including Target and Amazon to make the most out of whatever data about you they can get. In the hyper-personalized retail environment, your likes, dislikes, ZIP code, income, habits, gullibility, and friendships will one day affect not just your impulse buys and online shopping, but every purchase you make at the gas station, the coffee shop, and the grocery store, even including the price you pay for tuna.

This is one of the areas of predictive analytics where Gryc sees opportunity and change. "Right now, the corporations can afford these analytics. They can afford the data. But what's going to happen in the next few years is that it will become a lot easier for consumers to calculate a lot of these metrics."

The naked future envisioned by Gryc is one in which consumers and retailers are locked in an information arms race and he's the arms dealer. Today, one side has a clear and seemingly insurmountable advantage. But consumers have more information at their disposal than many realize. Any bank or wallet app can monitor in real time the amount of money we spend on things we don't really need and may not even want. A number of other apps can help you discover the consequences that a given purchase will have on your waistline, bank balance, or your goals. An app called Oroeco can even help you track and predict the effects of your purchases on the environment. Technology helps companies better predict where we'll be, what we'll buy, what we'll want, but it also helps consumers consume smarter, and sometimes even consume less. The big data present is one where companies use our data against us, to trick, coerce, and make inferences that benefit them at our expense. That behavior won't change in the future, but with a better awareness of what's going on and a willingness to experiment with the right tools we can make the fight a bit fairer. With enough personal record keeping, it's possible to turn the tables on the ever more coercive advertisers. For instance, using a QS system such as the earlier mentioned Tictrac, you can see how the media you expose yourself to affects your purchasing behavior, your ability to meet your own savings goals. Indeed, you can see how your exposure to Facebook changes your happiness and your financial security.

You have all the information that you need to help you resist ever more coercive mobile messaging; you give it away to your phone all the time. The next step is to start using it, to become smarter about you. Imagine answering a push notification on your mobile device and seeing the following message:

There is an 80 percent probability you will regret this purchase.

The answer to highly customized, context-aware advertisements is the strategic, personal use of personal information. The war is just beginning. Both sides will experience victories and perhaps moments of true partnership.

But not every party will emerge from this war as a winner.

Back at Saatchi & Saatchi, there's a clamber of electric saws outside. Down the hall, long slabs of drywall sit against bare beams. Carpenter stations are set up. I ask Becky Wang if the company is expanding. She tells me that the offices are shrinking. "We lost some business," she explains. One of the workmen stops by her office. "Hi!" she says. "I've been looking for you. I've been trying to use the computer in the hallway, but someone keeps stealing the keyboard. Can we move it?"

As we walk to the elevator, Becky gives me the names of people in New York and Silicon Valley whom she considers to be leaders in retail analytics. They work out of tiny start-ups that I've never heard of. The conversation feels a bit melancholy. The company where she works, with its carefully constructed facade of cool invincibility, is vanishing piece by piece. Before the elevator opens, I ask where I can follow up.

"Use my Gmail," she says and then, quietly, as though not to alarm the people around her that the vessel they are on is flooding and she is stepping into the last lifeboat, "I'll be leaving here, too."

CHAPTER 7
Relearning How to Learn

THE year is 2020. You're at a parent-teacher conference on the eve of the first day of a new school year. Your daughter is going into freshmen algebra tomorrow and you're at this conference to meet her new teacher. You show up armed with every math quiz, every math *problem* that your child has attempted throughout elementary and middle school, as well as a breakdown of how long she took on each and at what time during the day—after breakfast, before dinner—she performed best. The profile may even reveal whom your daughter talks to online, whom she studies with, and how those supposed friends influence her homework performance. This is a lot of information to carry around. If you were to print all this material, you would be dragging boxes along behind you. But this information is already stored on the cloud. All you have to do is give your child's teacher a link.

You have a request: "Would you mind taking all this data and creating an individual learning program for my daughter to make *positively* sure she finishes this year with an understanding of algebra? By the way, she's very shy, won't ask any questions in class, and probably can't devote more than an hour to algebra a night. Thank you."

Eight years of quiz scores footnoted and time-stamped? Facebook friends? Television-watching habits? What teacher in 2014 has the time to figure out the relevance of all that information? Not when lesson plans need writing, parents need to be called, and quizzes need grading. Thankfully, this isn't 2014.

Your daughter's teacher opens the link on her phone and downloads the relevant files. They're automatically run through a modeling app that sends her a notification. She suddenly knows exactly how well your kid will do on the first four quizzes, right down to which errors she's going to make. "It seems your daughter keeps reversing second- and third-order operations. We'll start drilling on those tomorrow. I'll schedule a half-hour online tutoring session for the evening, right after *Teen Mom*?"

"That would be great," you answer, thankful that you aren't being asked to come between your daughter and your daughter's violent devotion to her favorite show on TV.

"There's one more thing," says the teacher. "The profile shows that when Becky is confronted by a particularly hard problem she'll switch over to Drawsomewords 8 for five minutes or so. She seems to have great spatial-representation skills. I have a friend that designs drafting freeware at a studio downtown. I think that if we can get your daughter an internship, it might help her make the connection between math and drawing and then she'll exhibit a bit less resistance to second- and third-order operations."

This offer seems generous, perhaps too much so. "Isn't she a bit young for an internship? I mean, it's her first year of high school."

The teacher nods politely. "She's a bit late, actually. The average student her age has already started a company. But I think I can pull some strings."

HOW does the above scenario become reality (and do we want it to)? For starters, the feedback loop between a teacher administering a lesson and a student taking a test needs to collapse to the second or two it takes a student to click a mouse. More important, the time and convenience costs of keeping records on individual student performance would need to fall to virtually zero. Finally, teachers, state education secretaries, administrators, parents, and employers would have to be willing to accept new performance metrics in place of what we today call grades. Every item on that list, except for the last one, exists in 2014.

But the most important step is philosophical. We need to acknowledge what education is today: essential, expensive, and in terrible shape. The United States spends more than $10,000 a year per elementary and secondary student; that's $2,000 above Japan and $4,000 above South Korea, two countries where students are outperforming us in science and math.

Even if we don't know how to invest in school, we understand its importance. We've absorbed the fact that high school should prepare students for college because a college degree has never been a more essential credential to join the middle class. People with different education levels experience the same national economy in dramatically different ways. Unemployment among people with a high school degree was 8 percent in December 2012. Among people with a bachelor's degree it was 4 percent. Statistically, people with a master's degree or higher saw no employment collapse during the Great Recession. While it's true that nearly half of all 2012 college grads in the United States were either unemployed or, far more likely, underemployed in low-wage jobs (and carrying an average of $27,000 in school debt), they were still faring better than their peers who did not have a degree.

This speaks to a national skills gap that's growing along the lines of economic class. Low-skilled jobs are partly being replaced by a smaller number of high-skilled jobs. Even as GM parts factories were shuttering in Michigan, kids in Silicon Valley were seeing their start-ups bought out in a matter of months. In many cases it wasn't because of the products or services those fledgling companies were building but because of the talent contained therein, a phenomenon sometimes called acqui-hiring.

Our nation's response to our education challenge (both locally and nationally) embodies the worst aspects of an obsolete mindset. A slavish devotion to lecturing has been compounded by a nascent obsession with testing. Whether it's the Adequate Yearly Progress (AYP) reports mandated by No Child Left Behind, SAT scores, or just finals, the effect is the same: at the end of a designated interval—a week, a semester—teachers ask students to take a test. Too often we accept whatever result comes back as an objective and useful indication of the students' command of the material (administered via lecture). We do this despite the fact that history is full of intelligent people who didn't perform well on standardized tests and we know people forget information they've been successfully tested on. A lot of this testing is purely for the sake of identifying failing schools and teachers. Increasingly little of it has to do with helping students learn. Lectures make testing necessary. Testing makes lectures important. Testing is the big data present.

The naked future looks very different.

The Teacher as Superstar

The year is 2007 and Stanford professor Andrew Ng is in front of four hundred students, giving his famous and highly rated lecture on machine learning. He asks a question of the undergrads assembled before him and observes three distinct behaviors in response.

Ten percent of the class is slumped back; these students are texting, checking Facebook, or recovering from hangovers. They're what you might call "zoned out." About half the students are still madly typing the last thing said, displaying the sort of dedicated academic seriousness that propelled them through AP courses to get to Stanford. But they aren't raising their hands. Thirty percent or so sit quietly, waiting for someone else to answer. Only a few kids near the front, less than 1 percent by Ng's estimation, ask to be called on. If one of them gets the question right, Ng can breathe a sigh of relief and move on to the rest of the material.

The predictable dreariness of this lecture hall exchange began to depress Ng. It's a scene you could find in virtually any lecture hall today. Indeed, the *lecture* has changed relatively little from the time of Socrates, as evinced by the fact that Plato spends most of *The Republic* following Socrates around taking notes. It's a method of teaching that has endured because it's functional, which is not exactly a compliment.

When Ng looked out over that horde of four hundred students, he recognized himself among them, one of the quiet kids, neither waving his hand nor asleep, simply sitting, passive and indecipherable.

"I was a shy kid back in school. So raising your hand and asking a question, or answering a question, I did that sometimes, but not always," he tells me in his office on the Stanford campus.

Andrew Ng, it turns out, was fortunate to be a quiet student. If not for this quality of bashfulness, he would never have started his company, Coursera, which is remaking education for the twenty-first century.

Today, anyone in the world can familiarize themselves with the fundamentals of machine learning through Andrew Ng's massively open online course (MOOC). It boasted more than one hundred thousand alumni by July 2012. In his interactive instructional videos, Ng comes across very much as he does in real life. He is polite, serious, attentive, constrained in his movements, but friendly. He is not as shy as he was as an undergraduate at Carnegie Mellon but he remains an *exceptionally* soft-spoken man. Though he lectures quite successfully to auditoriums, he is clearly an instructor who thrives on one-on-one exchanges. His online course affords him the opportunity for this type of interaction with tens of thousands of people.

Coursera offers a huge departure in the way student performance is measured and understood. Instead of tests at the end of the week or semester, short, interactive quizzes are interspersed throughout the lesson, in keeping with the human attention span as science actually understands it (not how headmasters want it to be). Every student must interact with the material as they're studying it, not afterward. This allows Ng's online platform to be not only an information distribution system but a telemetric data collection system.

"We can log every mouse click, every time you speed up or slow down the video, every time you replay a particular five-second piece of the video. Every quiz submission, be it right or wrong, we know exactly how many seconds you took to do every quiz, and every post you read or posted. We're starting to look at this data, which is giving us, I think, a new window into human learning," Ng told me.

He admits that the subject matter in his machine-learning course is not easy. In fact, without a good understanding of linear algebra and at least some familiarity with statistics, the course is impossible. Chris Wilson from the online magazine *Slate* attempted the course and noted despairingly, "Avert your eyes, Mom, because I have a confession to make: I'm not entirely certain I'm going to pass."

Writing code for learning algorithms doesn't become intuitive just because we want it to, or because the White House has a renewed interest in science, technology, engineering, and mathematics (STEM) education, or because someone designed a video game to teach it. Computer science will remain a difficult, multistep, and rule-filled domain because such is science. Though we are prone in the Internet era to lionize technology wizards the way we used to venerate rock stars, science and music aren't interchangeable. Science will never feel natural because it is not natural. For all his genius, Andrew Ng can't change this.

What telemetric education offers is the chance for all students to raise their hands and be heard. That opportunity doesn't come easily in a crowded classroom and especially not for women or minority students, many of whom feel that if they ask the wrong question or display ignorance, they'll confirm some unflattering, broadly held perception about their social group. We now understand this to be a real phenomenon, one that plays out in classrooms around the world every day, called stereotype threat.

It turns out other people's bad expectations are holding *you* back.

Here's how we know this is true. In 2006 Smith psychology researcher Maryjane Wraga and a few colleagues gathered together fifty-four female students and paid them $20 apiece to perform a series of spatial tests. Wraga divided them into three groups and told the first group that these were the sorts of tests women were expected to do well on and then told the second group that they were *not* expected to do well. In essence, two-thirds of the participants were told that they were going to confirm or refute either a positive or negative stereotype about all women. She didn't tell the third group (the control group) anything.

Each subject took the test under functional magnetic resonance imaging (fMRI). The women who were told they were being examined to confirm a negative stereotype showed activity in the part of the brain associated with processing anger and sadness (the rostral-ventral anterior cingulate) and the part of the brain charged with learning about social and interpersonal relationships (the right orbital gyrus). In other words, the subjects themselves encoded the stated premise of the experiment—that women were more likely to perform negatively on the test—as fact.

Conversely, the second group of women, the ones who were told that the test was intended to validate a positive stereotype about their sex, displayed activity in the portion of the brain associated with working memory (right anterior prefrontal cortex) as well as the portion of the brain associated with egocentric encoding (middle temporal gyrus), which is how we perceive objects in relation to us.

We've known that confidence can affect test performance but until Wraga's study, science didn't know exactly how large a role social stigma and stereotyping play in education. Wraga and her colleagues found that the women with the positive stereotype stimulus did *14 percent better* on the challenge than the women with negative stereotype stimulus.[1]

Stereotype threat could be a contributing factor in the fact that just 30 percent of African Americans and fewer than 20 percent of Latinos have an associate's degree (among those currently in their twenties). It may also be one of the reasons why the United States now has a higher college dropout rate than any highly developed country.[2]

Consider the implications of Wraga's findings, particularly that 14 percent performance differential between the two subject groups. Bad performance doesn't result in stereotyping; rather, the situation is reversed. When a person is continuously exposed to negative predictions about how she'll perform on a test as a factor of group affiliation it's the *prediction* that has a deleterious effect on performance. Coursera creates an environment where students are shielded from the effects of these predictions.

So far, the system's biggest asset has been a collapse in the cost to do certain types of education and curriculum experimentation. "In a traditional education study, you may have twenty students in your experiment group and twenty students in your control group. And if you're really lucky, maybe you get something that's just barely statistically significant," says Ng. An extremely large education study, encompassing on the order of hundreds of students, can cost thousands of dollars and sometimes won't produce actionable results for months or years. On Coursera, every interaction can become an A/B test in which one-half of the test-taking population is shown one lesson and the other half is shown a different one. On any given day Ng can run such a test on twenty thousand students. All of this data is helping him understand the process of learning in a way that is specific to any individual and yet broad enough to be applicable to any student in any country.

In 2012, when two thousand of his students submitted the same wrong answer to a question, Ng realized immediately the problem wasn't the kids taking the course. Rather, the malfunction lay in the question itself: "In a normal class, if two students out of a hundred submit the same wrong answer, you probably won't even notice. But when two thousand out of a hundred thousand students submit the same wrong answer, that's a very strong signal to the instructor that some clarification is needed." Ng checked the question again and at first didn't see any error. Most of the students got it right. But when he applied a learning algorithm to the data set, he discovered that the wrong students were all making the same type of error: they were reversing two steps in the formula. In effect, he learned to predict errors before they occurred. He created a customized error message so now every student who misses that question is given a clue, a message urging him or her to go back and reconsider the order of the operation, which allows the students to correct missteps and move on much more quickly.

These sorts of discoveries are increasingly common in online environments that actually collect and use student data. When MIT physicist David Pritchard first analyzed the results of a widely used concept test given to about a thousand students, he found that most of the carefully designed problems tended to challenge the problem students but were a breeze for the better-prepared kids in the class, as anyone might have expected. But two problems in particular produced a counterintuitive result. The A students did well and the B students did less well; however, the C and especially the D students did better than the B students! He realized that the wording of the question was ambiguous, but the A students seemed to know what was being asked. Many of the students doing less well, after misunderstanding the question, had a common physical misconception that resulted in the correct answer—two errors combined had a canceling effect. It's the sort of event that a regular teacher lecturing to a class of thirty would never notice but because Pritchard had hundreds taking the concept test, the size of the data set enabled analysis that made the error visible.

Rapidly adjusting lessons on the basis of new incoming information is only possible in the Internet era. But Ng is careful to point out that he's not actually an advocate of Web-only education. He's a fan of what's sometimes called the flip model, which is in-person education with a heavy online component.

Far more students today have seen Ng's lectures because of the popularity of the machine-learning course, but he actually spends much less time lecturing than he used to. He uses his class time to teach high-level concepts, try out new material, and workshop big problems. "The instructor can look at a Web site and see what the students are getting right and getting wrong, so that [the instructor] can focus the classroom discussion on what the students are actually confused about."

Some of the most pioneering work in this field was conducted by Kansas State University physicists Dean Zollman and Sanjay Rebello, who handed PDAs to their students in 2005 and instructed them to actually text in class (so long as the subject was physics). Zollman and Rebello were able to quiz every student in real time and then alter lesson plans accordingly.

In many ways the Rebello-Zollman classroom provided a snapshot of what Ng is doing, and, perhaps, all classrooms of the future. In a press release Rebello remarked that the system worked well to address issues of minority student engagement. "I find that even in a small class it can give me a feeling for how this silent majority or silent minority of students is thinking about things that I wouldn't [normally] get."

Ng already has a lot of competition in the online learning space. On the East Coast the edX program, which features online interactive courses from MIT and Ivy League schools, has already attracted thousands of participants. On the West Coast, two of Ng's colleagues at Stanford, Sebastian Thrun and Peter Norvig, put their own interactive course in artificial intelligence online at around the same time and are now spearheading a company called Udacity.

Norvig also serves as a director of research at Google and is one of the most senior executives of the demographically young company. Prior to joining the search giant he led the computer science division at the NASA Ames Research Center where he worked on sending robots into space. He's considered one of the world's top minds in AI but is also known for his oddball sense of style. If you catch Norvig at a big event, he'll likely be in one of his trademark Hawaiian shirts. In public he speaks in a slow and swampy-deep voice that seems to emanate from the very bottom of his six-foot-and-then-some frame. The mathematical genius and the inner joker seem to have reached a strange but solid equilibrium. In teaching his online course, he found that little mistakes and moments of goofiness that can screw up a live lecture actually work well online.

"I thought it was about recording videos and making them flawless. What I discovered is that it's about making a personal connection. Finding a way for the students to make a commitment to do the work," he told me.

I first met Norvig at the 2007 Singularity Summit in San Francisco. We chatted for a bit and the subject soon turned to education. At the time I was needlessly concerned about the effect that Google could have on the future of learning. I pointed out that 50 percent of high school seniors (at the time) couldn't tell the difference between an objective Web site and a biased source. I asked him what he would do to preserve critical thinking skills in an era when such technologies of convenience as Google seemed to be doing a lot more "thinking" and students a lot less.

He answered, "What can you do about that? I think part of it is education. We're used to teaching reading, writing, and arithmetic; now we should be teaching these evaluation skills in school and so on. Some of it could be just-in-time education. Search engines themselves should be providing clues for this."

When I ran into him again at the 2012 Singularity Summit he was about ten pounds lighter and still wearing his trademark Hawaiian shirt. He was well on his way toward fulfilling the forecast he had made several years earlier: using the technologies of data storage, retrieval, and machine learning to bring just-in-time education closer to reality for millions of students.

"We're going to make rapid advances in understanding what works and what doesn't on the basis of interaction statistics," he told me. I asked him if he believed every student would have access to a predictive model of their own learning style in the next five years; if, in effect, there was a naked future for education. "That's the hope," he answered.

Today, online learning platforms that measure individual student performance are in their infancy. Andrew Ng, Sebastian Thrun, and Peter Norvig are still learning what works and what doesn't. But it is within the capability of the Coursera platform to offer individualized instruction on a level that surpasses what almost any in-person teacher today provides.

"We are analyzing the data as we try to understand when students are likely to have problems, when they are at risk for dropping out of the course. These are things we're looking at. And eventually, hopefully, this will allow us to catch and to encourage the students along when they need that bit of the extra encouragement. These are things we're working on. It hasn't happened yet," says Ng.

So far in this book I've covered some of the ways we've begun to accept rapidly evolving notions of privacy. In such areas as epidemiology, sharing more personal data can have a positive effect but at the personal cost of telling the public that you're sick. It's nakedness in the worst sense: the benefits are public and the costs are private. Learning and education is one example of where individuals will begin to see benefits of exposing a bit more of their life stream. Those who elect to take a different course will find themselves ever more at a disadvantage.

We haven't begun to understand how those depreciating costs of collecting and analyzing real-time scores, footnotes, and time-stamped marginalia will change the definitions of teacher, student, and learning. If Andrew Ng and Peter Norvig's experience is any indication, teachers are going to realize the dual and compounding benefits of being able to provide much more personal attention to a far greater number of students. The teachers who are extremely effective at this will model the way forward for everyone else. But in the coming decade, teaching will come to mean something very different than what it means today.

The End of Teaching

It's January 2012, the setting is the Ethiopian village of Wenchi, which sits on the rim of a volcanic crater lake some eleven thousand feet above sea level. Wenchi village is one of the poorest communities in one of the poorest countries in the world. The majority of the houses here, without running water or electricity, are scarcely larger than a single room. They have dirt floors and roofs constructed of branches, which seem to provide little more than shade.

A group of non-governmental organization (NGO) workers arrives in the village from the nearby capital of Addis Ababa. They bring with them a sealed cardboard box. The workers have a brief conversation with one of the village elders and explain that inside the box are tablet computers, one for every child in the village ages four to eleven. The tablets have been outfitted with solar screens so they can charge in the sun and come fully loaded with hundreds of apps, movies, and games, all in English. English is the official language of this village and this was the chief criterion in its selection for this gift. But "English speaking" is just a technical designation because the population of this village is illiterate down to the last person and unable to understand instructions or subtitles that are part of the apps or the movies on the devices. In fact, almost no one here has ever encountered the written word in any form. There are no street signs, no candy wrappers, Coke bottles, flyers, or advertisements. It is a tabula rasa in the desert.

The relief workers from Addis Ababa leave the box and return to the capital. They will circle back to the village soon, in about a week or so, to swap out the subscriber identity module (SIM) cards in the tablet PCs.

Software running on the tablets will log every keystroke and swipe when it occurs. The devices will record each child's progression through the games and apps, step-by-step, command by command. The cards will then be express mailed from Addis Ababa to Cambridge, Massachusetts, where Nicholas Negroponte and a team of researchers he's assembled will analyze the data and search for clues to reveal how the children are improving at written and spoken English despite having no instruction whatsoever.

Skip ahead a few months. On October 25, 2012, Negroponte stands in front of a gathering of technology enthusiasts at MIT to relay what months of research have uncovered. "In five days they were using forty-seven apps per day," says Negroponte. "In two weeks they were singing ABC songs." By the end of five weeks, Negroponte reports that several were able to activate the tablet's disabled camera and take their photos. "They hacked Android!" he shouts to the crowd.[3]

Negroponte is the founder of the MIT Media Lab. He's credited with being the first investor in *Wired* magazine and is the author of the 1995 bestseller *Being Digital*, a forward-looking treatise on the future of man and machine that seems to grow only more influential with the spread of the Internet. He's probably best known for his One Laptop per Child (OLPC) Association, which donates specially designed laptops to children across the developing world.

When you go to the association's Web site, you'll see pictures taken of OLPC initiatives across the globe. One in particular has become closely associated with the program. It features several dozen Nigerian boys all about age ten. Each one is wearing a school uniform of avocado green with forest green trim. They sit in neat rows at long schoolhouse desks. Each boy has a bright white and green laptop. These devices, set against the blandness of the cement wall classroom, the rustic benchlike tables, the dour green uniforms, look like artifacts from the future deposited in the past. All of the boys are smiling politely.

A casual observer would say the picture speaks to the enormous success of the program, which has given away more than 2 million laptops around the world since the first full model was revealed in 2005. But Negroponte admits to a degree of ambivalence about this scenario.

"This was the image that started to spread because, sadly, our partner was always the government, and the government wanted us to do schools." He says "schools" with a distaste that's reminiscent of a ten-year-old at the end of summer vacation.

Negroponte doesn't measure the success of the program on the basis of units distributed or test score improvements. Not only are these the wrong metrics but he believes the continuous focus on testing, curriculum, indeed on *teaching*, often gets in the way of real *learning*. The assumption that the latter is impossible without the former is wrong. "We know that a vast recall of facts about something is in no way a measure of understanding them," Negroponte wrote for the online edition of *MIT Technology Review* just before his lecture. "At best, it is necessary but not sufficient. And yet we subject our kids to memorizing. We seem to believe that rote learning is akin to physical exercise, good for their minds. And, quite conveniently, we can test whether the facts stuck, like spaghetti to a wall. In some cases knowledge is so drilled in that you know and hate a subject at the same time."[4] For Negroponte most of what we call school is a gratuitous time suck at best, and a real impediment to learning at worst.

This view is somewhat unusual for someone who is, at least in name, an educator. But Negroponte isn't the first teacher to condemn the institutionalization of education, which has been an evolving process throughout history. In his writings and speeches, educator John Taylor Gatto describes the emergence of a teaching system that, from the 1800s onward, served primarily as a means to house the children of factory workers while their parents toiled in sewing rooms, slaughterhouses, and mills. The objective of public education was to create more of the same. By its very nature, Gatto observes, formal education works to suppress creativity and unnaturally lengthen—as well as dull—childhood. "Mandatory education serves children only incidentally; its real purpose is to turn them into servants . . . After a long life, and thirty years in the public school trenches, I've concluded that genius is as common as dirt. We suppress our genius only because we haven't yet figured out how to manage a population of educated men and women. The solution, I think, is simple and glorious. Let them manage themselves."[5]

Negroponte's skepticism about school also recalls some of the more provocative insights of the British philosopher John Stuart Mill, who observed, "A general State education is a mere contrivance for molding people to be exactly like one another; and as the mold in which it casts them is that which pleases the dominant power in the government, whether this be a monarch, an aristocracy, or a majority of the existing generation; in proportion as it is efficient and successful, it establishes a despotism over the mind, leading by a natural tendency to one over the body."

The MIT Media Lab that Negroponte cofounded exists in stark contrast to the picture of education painted by Mill. Classes happen but the emphasis is on cobbling and tinkering. "We don't teach at the MIT Media Lab. We do research. We assemble teams. We don't have students so much as apprentices," says Negroponte. His poor opinion of regimented schooling is, in part, why he created the OLPC Association. He had hopes that the program would decouple learning from teaching in the developing world just as the MIT Media Lab was helping young designers, inventors, and entrepreneurs build expertise outside the formal teacher-student relationship. He was soon deflated to realize that this emphasis on rote memorization, testing, discipline, and regimentation was even more pronounced in the developing world than it was in the United States.

In his lecture at MIT he recalled going to villages in Pakistan and meeting first-graders who were actually excited, "wide-eyed," he described them, for the first day of school. He returned two years later and discovered those same children, who were then third-graders, as we encounter third-graders today: subdued, uninterested, and robbed of the desire to learn.

It was on one such trip while he was in nearby India that he met Sugata Mitra, a computer scientist with a PhD in the theoretical solid state physics of organic semiconductors from Indian Institute of Technology Delhi. As a physicist Mitra had an interest in complex ordered systems in which the actors organized themselves into coherent forms without outside intervention, similar to bee and ant colonies. Mitra believed that human learning and knowledge formation might be subject to the same invisible, cohesive forces. But how was he to test this idea, as any human-led effort to set the parameters for knowledge formation in a subject (in this case, a child) might be called teaching?

Mitra's solution was to go around the human.

In 1999, at his office in New Delhi, he launched an experiment he called the "hole in the wall" to demonstrate that self-organized learning via an electronic interface—with no teacher intervention at all—was possible.[6]

He and his team members "cut a hole inside that wall and put a pretty powerful PC into that hole, sort of embedded into the wall so that its monitor was sticking out at the other end," Mitra explained at the 2007 Lift Conference in Geneva, Switzerland. The PC had a touch pad and Internet.

A few hours later, an eight-year-old boy, shoeless and dressed only in a dirty kurta (long shirt) from the nearby slum neighborhood (Mitra's description) of Kalkaji, approached the hole and began to play with the keyboard and the mouse. The boy had never seen a computer before. A few hours later, when his six-year-old sister showed up, the boy was able to tell her—without knowing exactly what he was talking about—how to browse on AltaVista.

Mitra performed the experiment several times across India. In a rural village called Madan Tusi there was no formal English instruction at all. He set up a hole-in-the-wall kiosk, left behind some educational CDs in English (the village was devoid of Internet service), and returned a few months later. "I found these two kids, eight- and twelve-year-olds, who were playing a game on the computer. And as soon as they saw me they said, 'We need a faster processor and a better mouse,'" Mitra told the Swiss crowd.

When he surveyed the kids, he found they were using an average of two hundred English words in casual communication with each other. Though they couldn't pronounce the English correctly, they understood the proper usage for each of the words. In other cases, where the hole-in-the-wall computer was connected to the Internet, the children taught themselves how to browse and e-mail, how to use drawing applications and the most basic Windows functions. But this learning only took place when a big group of kids was crowded around the computer, jostling and shouting and playing, as kids do.

This is an important point. We associate educational technology with lots of individual screen time, such as those kids at desks in forest green uniforms, each kid burrowed into his own laptop. The kids in Mitra's experiments spent far less time learning from the machine than from one another.[7] Mitra realized a startling conclusion.

A machine doesn't teach you as well as the act of teaching others.

"What you have, actually, is there is one child operating the computer. And surrounding him are usually three other children, who are advising him on what to do. If you test them, all four will get the same scores in whatever you ask them. Around these four are usually a group of about sixteen children, who are also advising, usually wrongly, about everything that's going on, on the computer. And all of them also will clear a test given on that subject. So they are learning as much by watching as they learn by doing," Mitra explained to the Lift group.

In the OLPC program, Negroponte observed something similar. He estimates that of the 2.4 million cases where laptops were distributed, in 3 or 4 percent of the cases the kids taught their parents how to read and write.

The findings are inspiring but not particularly conclusive. They don't suffice as an argument for radically redefining primary school education. Mitra could only observe the aftereffects of putting the computer in the town square, just as Negroponte can't stand over the shoulder of 2.4 million kids to whom he's given laptops in order to observe how they learn and when.

But Negroponte's Ethiopian tablet initiative takes an important step forward; it closes that feedback gap in the way it records every interaction that takes place over the medium. Though not exactly telemetric, this data could provide clues as to the causal link between group interaction and learning. Those clues exist in a time between the two moments—months apart—when the cardboard box was dropped off in Wonchi village and when the kids hacked Android. The data are open to anyone.

You might describe the one tablet experiment as exploitative.

It's hard to imagine an American parent reacting well to the idea of her child coming home from school with a "free device" loaded with shadowy surveillance software, even if it is part of an experiment that comports with International Review Board standards. Many others have criticized the entire OLPC project as expressing too much faith in the Western technologies of personal electronics, particularly the emphasis on English. In 2005 computer designer Lee Felsenstein likened the initiative to dropping flat-screen TVs on villagers from helicopters: "It is sufficiently discomfiting to consider that the outcome of a massive project like OLPC may be a different form of commercial television for the developing countries. Worse yet would be the preemption of funding for many other projects designed under a community model. Future talk of computer systems for the developing world would meet the dismissive response that 'it's been tried and it failed.' "[8,9]

As an act of charity, dropping computers on the world's poorest kids is self-serving and patronizing.

This is one view.

An opposing view holds that Mitra's and Negroponte's work is a long-overdue response to one of the more enduring legacies of *actual* colonialism: the continued export of the classroom model to ever-poorer communities where the model does not always work. In separate research conducted throughout India, Mitra has found that the quality of education declines as a function of distance from the capital. Teachers in the more remote villages simply lacked either the desire or the ability to perform at the same level as their urban counterparts.[10]

Mitra doesn't share Negroponte's animus against all schooling but he does argue that educational technology programs are implemented in a way that protects the status quo first and foremost, as new ed tech is piloted in schools that are already well funded and where students are high performing. Poorer or remote schools where educational technology can do the greatest amount of good don't get the same opportunities to try out new solutions.

As Mitra told the Swiss crowd, a computer can't replace a good teacher, but if it can replace a *poor* teacher, then we should let that happen. "I'm proposing an alternative primary education is required where schools do not exist, where schools are not good enough, where teachers are not available, or where teachers are not good enough."

This simple plea doesn't sound mutinous but it is. Telemetric platforms and technology hold the potential for a change in education that will resemble revolution. When metrics are no longer constrained by the time it takes to grade paper tests, *any* metric could fall by the wayside and be replaced by new signals that are more truthful.

For instance, Laura Matzen of Sandia National Laboratories and some of her colleagues have demonstrated that the brain's electrical activity, detectable via electroencephalogram (EEG), predicts how well-studied material has been incorporated into memory, thus how well a subject will perform on memory tests.

The researchers asked twenty-three people to attempt to memorize a list of words while undergoing a brain scan. The average subject recalled 45 percent of the words on the list. The EEG data correctly predicted which five of the twenty-three subjects would remember 72 percent of the words, beating the average.[11]

"If you had someone learning new material and you were recording the EEG, you might be able to tell them, 'You're going to forget this, you should study this again,' or tell them, 'Okay, you got it and go on to the next thing,'" Matzen said in a press release.[12]

Imagine for a moment the power of knowing beforehand how well you would perform on a test but how disempowered you would feel if that same future was naked to your competition, or to your future potential employers.

Telemetrically gathered scores aren't a perfect indication of future performance, so an inherent danger exists in relying too much on them. Likewise, educators, and particularly school districts, should avoid the temptation to rely on MOOCs to quickly solve education disparities that have been building up for decades. In 2013 San Jose State University (SJSU) entered into a deal with Sebastian Thrun's

Udacity to allow the online platform to provide remedial instruction in certain classes. SJSU has a lot of students who need basic training in math and language before they can take regular college-level courses. To some, the deal looked like a cheap way of getting around the fact that public primary and secondary education in California is anything but equal. On the *Atlantic* blog Ian Bogost mockingly summarized the deal: "The answer to underfunded, lower effectiveness primary and secondary education requires subsidizing a private, VC-funded bet made on a roulette wheel fashioned from the already precarious prospects of a disadvantaged population."[13] Other critics point out that the primary user group for MOOCs isn't kids who are having a hard time getting into higher education but students who have already completed two years of college. Some members of the online education community with whom I've spoken are quietly aghast at the idea of replacing remedial classes with MOOCs. They feel that the real potential of these education platforms is in adult and continuing education. The Udacity-San Jose State online college partnership has been marked by low completion rates for the course among students who need it most. Thrun has since suggested that Udacity will "pivot" away from general education toward corporate training. The critics clearly have a point.

Even if it's radically transformed, one reason school will probably endure is because the act of establishing interpersonal connections can't easily be digitized. For this you need co-location, a setting in the real world.

Andrew Ng began his experiment in online education with his colleague Daphne Kohler. It was an experiment that began simply, a few lectures posted to YouTube, no feedback, no two-way collaboration, and no interaction. It was the standard VHS lecture but with a cheaper look and on a cheaper medium. By Ng's account it didn't result in much and that's how history would have left it had he not had the fortune to be teaching machine learning in the heart of Silicon Valley where there are a lot of people with a keen interest in such obscure technical subjects as machine learning. There are students who match this description in Oklahoma City, or Cleveland, or Lyon, France, but you don't run into them with the frequency that you do in Silicon Valley.

Shortly after putting his lectures online, Ng kept running into folks who had taken his course. He knew, from YouTube, that tens of thousands of people had seen the videos but the experience of encountering random strangers, students who appreciated the effort, who wanted more, turned out to be critical for reinforcement.

"I was fortunate," he acknowledges. "It's a big circle."

A common complaint of the Internet era is that the spread of information technology inevitably creates distance. Some, like Professor Michael Bugeja, have suggested that the withering of interpersonal connections is the inevitable result of the creeping digitization of life. The growth in the number of conversations and exchanges that take place purely over digital media can surely contribute to feelings of alienation. But the stories of Andrew Ng and the MIT Media Lab suggest that certain communities and the places they are attached to, which urban studies theorist Richard Florida calls "creative class cities," will only grow more important in the future. The most recent data on college admissions suggest that, far from hurting applications to Stanford and MIT, putting courses online has attracted a lot more applicants to these schools. The bottom line that this trend suggests is that MOOCs necessarily repair inequalities in America's education system even if they only address portions of it, such as stereotype threat.

But even if MOOCs aren't the ultimate solution, they can certainly help us improve the way we learn. A moving score, in the form of a continuously updating education profile, is probably a better indication of your potential than a static one that reflects who you were or what you could do when you were sixteen, or where you completed four years of schooling when you were in your late teens and early twenties.

Learning will become easier and much more of it will happen outside school settings, all of which will diminish the importance of schools and teachers as we know them today. But platforms such as Coursera can amplify the talents of gifted and effective instructors and reduce the cost of education in the coming decade for all. *Some* schools and colleges will thrive and prosper at a level not seen in their history. But they will do so only by transitioning away from today's classroom model and toward something else, such as data-driven skills workshops at the high school level and start-up incubators or problem-solving workshops at the college level. The latter transition may be the hardest but it is also the most essential for the survival of higher education. Many less intellectually fecund colleges will find it hard to persuade young people to go into debt in order to get a credential that is increasing in cost and diminishing in value.

We may be conflicted about relying less on classrooms and more on platforms but if we are to be honest with ourselves, we know that we can't prepare coming generations for the challenges of a technological and globalized economy in the same way we prepared previous generations to be factory workers. The greatest thinker of the current century (a person whose identity we do not yet know) will understand more about how he or she thinks and learns than any student in any previous generation long before ever stepping foot inside a schoolhouse.

CHAPTER 8
When Your Phone Says You're in Love

MATHEMATICAL matchmaking sounds like a modern method for finding love. In truth, it predates dating Web sites, the computer, the field of statistics, or even *Europe*. The practice goes back to 1500 BC and the composition of the four Veda scriptures, part of the ancient Indian religion of Brahmin that later became Hindu. While the rules governing Western astrology are open, discoverable, and thus easy to ridicule, Vedic astrology as traditionally practiced was esoteric. Expertise was passed among a small, elite group with intimate knowledge of the ancient Veda texts. These Brahmins had a monopoly on the business of drafting personality profiles, or *janam kundalis*, which all well-to-do parents had to have before they could arrange marriages for their children.

Vedic astrology puts forward a mathematical explanation for human personality that has logical and weighted underpinnings, even if it is fabricated. Paramahansa Yogananda, whose *Autobiography of a Yogi* was one of the first accounts of the religion to be translated into English, explained the connection between an individual personality and planetary movement as a causal relationship.

"Astrology is the study of man's response to planetary stimuli. The stars have no conscious benevolence or animosity; they merely send forth positive and negative radiations. Of themselves, these do not help or harm humanity, but offer a lawful channel for the outward operation of cause-effect equilibriums which each man has set into motion in the past."[1]

Today, most of us in technologically advanced societies view arranged marriage as an inherently chauvinistic practice, seeing as its goal an exchange of daughters (mostly) for property or alliance formation and quite separated from anything resembling love. The history of the practice in the West would support that view. But *janam kundali*–based matchmaking in India, which is still practiced today, is somewhat less transactional. Compatibility is the goal. Vedic matchmakers consider dozens of elaborate weighted variables as part of a predictive model. It's believed that a promised pair should share at least eighteen matching points in their thirty-six-point *gun milan*, in addition to moon position. All of this formulation is based on careful observation of planetary movements. Today, many Indian couples who were married according to Vedic tradition swear by the practice and have long, stable marriages to bolster their case.[2]

That's not to say this method for pairing people holds up to any sort of serious scrutiny. The endurance of arranged Vedic marriages has nothing to do with moon position but with partner commonality. Social convention biases the results because throughout most of Indian history it was customary for individuals to marry within their caste. This ensured that couples had similar socioeconomic backgrounds and a wide web of mutual affiliations, two factors that have been proven to be predictors of marriage longevity. And, of course, divorce was not really an option.[3]

Today, we use better instruments and more complex math to map the movements of planets. While the field of physics has yet to reconcile the divergent equations of Sir Isaac Newton with those of Albert Einstein, it has given us a serviceable understanding of how matter and energy interact. There exists neither a plausible theoretical basis—nor any evidence—to support the idea that the position of the moon, the sun, or Neptune on the day of a person's birth will have any measurable effect on her personality, relationship, or the future.

In some ways, science has triumphed over superstition. But in many other crucial ways, the power of the esoteric has only grown. Humanity is even more open to the idea of matchmaking by math, *even math we don't understand*, than we were thousands of years ago. Many of the 40-million plus Americans who have tried online dating sites that use propriety (i.e., secret) matching algorithms aren't that different from the poor souls who paid hefty sums to Brahmin wise men for predictive models that couldn't be proven. To date, no online matching site has demonstrated the success of their algorithm in any way that would allow an independent skeptic to check their work or repeat their results.

The one online dating site that is relatively open about their matching methodology is the very popular OKCupid. The system uses two scores: how you answer questions and the importance you place on a potential mate's response to the same questions. The more questions you answer, the more information the system has to improve its matches. Users are presented with questions ranging from the humorous (e.g., "Have you ever murdered anyone?") to the personal (e.g., "How often are you open with your feelings?") to the basic (e.g., "Is the earth bigger than the sun?") to the political (e.g., "Is gay marriage a sin?"). The most provocative questions are extremely personal, dealing with sexual preferences, interests, boundaries, and history.

After a user responds to a question she is presented with an importance scale to categorize how she wants a partner to respond. Answers here range from "irrelevant" to "mandatory," indicating that the user does not want to be matched with people who avoid that question. The point value is logarithmic so the values increase exponentially as opposed to linearly.

How important is this question to you?

	POINT VALUE
Irrelevant	0
A little important	1
Somewhat important	10
Very important	50
Mandatory	250

(Source http://www.okcupid.com/help/match-percentages)

To calculate how well you match with someone else, the system takes your scores, the scores of the person whose profile you're viewing, multiplies these together, and takes the square root. If you're a 95 percent match with someone, that means you answered many of that person's *most important* questions "correctly" (i.e., in the way she indicated she wanted that question answered) and she answered your most important questions the same way. The logarithmic system ensures you're not matched with someone who just happened to share lots of trivial things in common with you but was miles away on the important stuff.[4]

Because users submit their own questions, there's a seemingly endless supply of them, and because the more questions you answer, the higher the likelihood of getting a good match, OKCupid is one of the very few Web start-ups outside health care that offers a real and tangible benefit for giving away more personal information.

Theoretically.

The success of this service illustrates how readily we'll give up extremely compromising information when asked. A great deal more tech coverage and scrutiny are given to Facebook for making minute changes in privacy policy than are given to sites like OKCupid, though the user-submitted material in OKCupid's databases makes the typical Friday-night Facebook post look positively Amish. When users tell their intimate secrets to OKCupid, they do so with an expectation of privacy. But the site reshares a surprisingly large amount of very sensitive information gleaned from questions such as drug habits and sexual orientation, with (at the time of this writing) nine different data resellers such as PubMatic, Lotame, Google's DoubleClick, Nexus, and Facebook. These outfits then go and sell that data to marketers looking to target customers.[5]

When OKCupid purges a user for violating the terms of service as happens from time to time, that user has no means of getting her questionnaire data out of the system or verifying that's it's been destroyed. If you leave OKCupid voluntarily, you don't get your data back, either.

In its early days, OKCupid was critical of such dating sites as Match.com, which charges a subscription fee of $35. Christian Rudder, one of OKCupid's founders, explained the start-up's philosophy in a 2010 blog post: "The practice of paying for dates on sites like Match.com and eHarmony is fundamentally broken."[6]

The primary malfunction was an imbalance of parties. As Rudder observed, "Men drive interactions in online dating. Our data suggest that men send nearly 4 times as many first messages as women and conduct about twice the match searches. Thus, to examine how the problem of ghost profiles affects the men on pay dating sites is to examine their effect on the whole system."[7]

OKCupid was eventually purchased by IAC/InterActiveCorp, the same company that owns Match.com, which also bought all of OKCupid's data. Sam Yagan, another cofounder of OKCupid, soon rose to become head of the entire portfolio of IAC dating sites including Match.com. He had a change of heart about paid dating.

"I think that there was a time where I believed that dating would be a winner-take-all market in the same way that Craigslist, eBay, PayPal, were all winners-take-all marketplaces. I think dating is different," he told me. Because Match.com has a much higher number of users across more age groups (93 million monthly site visits), there's a better chance of users finding a date, which is worth paying for, he says. More users also help both Match.com and OKCupid better understand the science of love.

In his first post on the blog, titled "Rape Fantasies and Hygiene by State," which showed a state-by-state breakdown of people who answered questions about their willingness to act out rape scenarios in bed at a partner's request, OKCupid cofounder Chris Coyne boasted about the utility of OKCupid as a living social science lab: "Old media could only get 3,050 people to answer a poll about Obama. And it was enough to call the election with confidence. OKCupid, on the other hand, can ask the world's most personal questions and get hundreds of thousands of answers."[8]

OKCupid, as a Web site, is indeed a provocative tool for measuring the attitudes, beliefs, and sexual peccadilloes of millions of people. But no matter how many questions users answer about themselves on the site, their matching percentages with other users aren't any real indication of how likely one or another of them is to enter a long-term relationship.

In a 2012 article from the journal *Psychological Science*, Northwestern University psychologist Eli J. Finkel and his colleagues show that at best, even very good online dating sites can help you rule out whom not to go on a date with, but can't tell you if the person you're on a date with is at all likely to become your lifelong partner. Likewise, it can't give you any advice on making that relationship stronger.

Here's why relying on profile compatibility alone doesn't work. When agents are free to select potential matches from a menu of profiles, certain profiles receive a lot more attention than others; and the more popular a person's profile, the more messages, chat requests, and invitations she receives, the less likely she is to answer *any* of them. The result is that the best candidates feel overwhelmed and don't want to participate in the network, and a lot of lesser candidates send out requests that aren't answered. Eventually, they lose interest, too.

You might call this the prettiest-girl-in-the-room syndrome. It's really just the social manifestation of survival of the fittest. From an evolutionary perspective, it makes perfect sense that we would constantly seek to associate with people who are a bit out of our league, even when sexual reproduction isn't an issue, but we don't want to be so outside our league that we give up any shot of scoring. Online dating sites mask the true odds.

Christian Rudder has acknowledged that the online environment makes the prettiest-girl-in-the-room syndrome worse. "You've got to make sure certain people don't get all the attention. In a bar, it's self-correcting. You see ten guys standing around one woman, maybe you don't walk over and try to introduce yourself. Online, people have no idea how 'surrounded' a person is."[9]

We experience online dating profiles in the same context we view Amazon items. We shop for what on paper looks best. But with online dating, there's an added illusion of one-on-one bonding. When we happen upon a profile that speaks to us we feel like we're getting to know someone for the first time, intimately. We can't see that dozens, perhaps hundreds of people are having the same reaction to the same profile. Dating Web sites would provide more value if they could predict which profiles were going to get the most attention. This is the score that really matters, and we've barely begun to understand how to tally it.

How Facebook Will Soon Be Able to Predict Whom You Will Like

In 1946 psychologist Fritz Heider first proposed a methodology, albeit a simple one, to quantify how what you liked affected your relationships, and how your relationships affected what you liked.[10]

His methodology, since dubbed balance theory, holds simply that when the people we like don't like the same things we like, we grow to either like those people less, tolerate their bizarre affections a bit more, or convince ourselves that the discrepancy is an illusion or irrelevant. Any one of the above options brings the relationship back into balance. And because balanced relationships require less energy to maintain, they are more sustainable. This is how he explained it: "p likes his children, people similar to him; p is uneasy if he has to live with people he does not like; p tends to imitate admired persons; p likes to think that loved persons are similar to him."

Balance theory, in other words, suggests that it's easy to be friends with those who are also friends with your friends, and your friends' enemies make suitable enemies for you as well. You could plot this out on a triangle with a series of pluses and minuses, and depending on how the pluses and minuses matched up, the relationship would be either balanced or imbalanced. In relationships that are out of balance, where you have two friends who dislike each other (and neither can pretend the other does not exist), you're very likely to drop the relationship that has the highest cost. That means that if you can accurately ascertain which of your friends like one another and how much each relationship "costs," you can predict which of your friends you will keep and drop.

In 2004 a computer scientist and one of the inventors of RSS, Ramanathan V. Guha, proposed an alternative—but still complementary—theory that status was a bigger factor than balance in who liked what.

Guha, who today works for Google, was the chief architect of Epinions.com, a product review site claiming to offer "real reviews by real people." In order to distinguish more trustworthy product reviews from less trustworthy ones, Guha created a "Web of trust" system where Epinions users could rate their fellow reviewers as authoritative or not authoritative. As tends to happen in online communities, some users quickly established more clout than others. Guha observed that those individuals who had the highest status had the most pluses attached to them (they were adored) and sent the most number of minuses out.[11]

Back to our problem. Status theory explains the prettiest-girl-in-the-room syndrome but balance theory much better represents what a functional relationship is like. So which theory works to predict romantic matches? The answer is both.

In 2010 Stanford's Jure Leskovec, who has done work with the Facebook Data Science Team, applied social balance theory and status theory to Epinions, Wikipedia, and Slashdot users to see which theory predicted how people would form alliances. Epinions and Slashdot allow users to designate "enemies" as well as friends, and Wikipedia allows users to edit the work of others (which can be a signal of an antagonistic relationship).

Using sixteen feature vectors, he found that he could predict friend and foe relations with up to 90 percent accuracy. Now apply this to Facebook, which doesn't allow "unfriendships" or "dislikes" but does allow comments on posts. Those comments that are not accompanied by likes can be correlated with dislikes. It wouldn't be hard to run comments through a semantic machine-learning algorithm to determine key dislike phrases that could more clearly indicate a negative edge (big minus sign). If you've got one friend who is constantly sharing material you aren't fond of—election season tends to bring this stuff out like nothing else can—and you find yourself arguing with her posts, there's a good chance you don't regard your friend as having a terribly high status; you perceive little downside to picking a fight with her. If you notice that one of your friend's friends tends to agree with you, there's a good chance you will wind up being that second friend's friend before too long (friends on Facebook, anyway).[12] Facebook, which also serves as a dating site for millions of users, gives a clearer window into status.

If you want a more precise understanding of someone's rank on Facebook, look beyond their friend count to the number of updates they post and the number of likes they get for them. While you may not have much of an interest in predicting which of your friends' friends you will connect with on Facebook, Facebook does have an interest here for reasons we'll get to later in this chapter.

Status and balance scores are what's missing from online dating sites, yet the reason for their absence is obvious. No one would use a dating site that made him feel like a loser with a terrible status score.

But Finkel's research shows that the long-term survivability of a romantic relationship is predictable on the basis of three variables. Similarity between partners, a category that includes music, religion, educational attainment, income, location, and a host of other things that can (today) be discovered online, is just the *first* one. The other two are how partners collaborate and interact on a day-by-day basis and how partners react to stressful events.[13]

You are more than a sexual fetish. For that matter, you're more than an income bracket, more than your last educational degree obtained, height-weight proportionality, facial feature symmetry, location, political affiliation, or musical taste. Most pay dating sites try to quantify your personality and some even try to give a number to how you react to different events. But they do this via a survey and that's the problem, because you're more than who you are when you sit down and fill out a form describing who you are. When you reduce yourself to a dating site profile, the result may be closer to the *ultimate you* than the position of the moon when you were born, but perhaps not by that much.

The second variable in predicting relationship longevity that Finkel identifies is collaboration style, a category including communication signals such as how well someone listens, how often or forcefully he interrupts people, and whether he laughs at his own jokes or never at all. Collaboration style includes subtle and nonverbal forms of communication: fidgeting, hand waving, posture, flirtatious glances, and disconcerting stares. These are factors that come into play when people talk about clicking with someone on a first or second date. But collaboration style also comes into play in working relationships and can include such factors as how likely someone is to ask for help when they need it; if she waits until the last possible moment to deliver uncomfortable news; if she seems to whine about every little thing. Whether the answers to these questions are deal breakers for a relationship depends on the unique nature of the couple and the way their communication influences each other. Opposites do sometimes attract because some communication styles have to be complementary, rather than reflective, in order to work. These are the sorts of qualities that simply don't make it into an online dating form, at least not yet. Measuring how two people communicate and how they collaborate has been historically extremely difficult.

That's beginning to change.

Let's Go Enjoy Karaoke

It is the month of April 1998. The setting is the Shibuya neighborhood of Tokyo, the most fashionable four blocks in the trendiest city in the world. Bright lights, two-story advertisements, and enormous television screens look down on the hip teenagers who are gathered below.

Kaori Mikuriya, sixteen, is hanging out with her friends. A signal chimes on her Lovegety, a small, pink, oval-shaped device she recently purchased for 2,900 yen ($25). The chiming indicates that a boy is within the device's fifteen-foot range. No doubt *his* device is going off, too, whoever *he* is. But who is he? She suddenly regrets ever buying this thing. She has no idea what sort of person her device may connect her with. "I started looking around while getting ready to run, if the boy was strange," Mikuriya later told *Wired* reporter Yukari Iwatani.[14]

The boy, it turns out, is not "strange." He's cute. Cute enough. They approach each other and green lights go off on both devices. They're a match, which is to say that they've put their devices on the same setting, of which there are three: "chat," "let's go enjoy karaoke," and "get2," which signals . . . an interest in something beyond karaoke. They adjourn to a *takoyaki* stand for a snack of breaded octopus tentacles.

Takeya Takafuji invented the Lovegety precisely to facilitate these sorts of exchanges because, as he explained to reporters, "Japanese men are very shy."[15]

Lovegety couldn't predict a perfect match, but it did boast one important feature that many dating Web sites still can't replicate: it forced people who knew almost nothing about each other to pick an activity and try it together, thereby affording shy Japanese men an opportunity to demonstrate that they could be fun activity partners.

Within four months of launch, Takafuji and his employer, German company Erfolg, had sold 350,000 units around the world.

One of them went to a young, single, British reporter named Charlotte Kemp who wrote of her experience for the *Daily Mirror*. She took the device to her local bar and found, unfortunately, that a matchmaking widget that worked well with shy Japanese boys in Shibuya had a very different effect in a pub on the East End of London. As soon as she turned on her device, the men in the room responded like sharks answering the smell of blood. "Men can be so obvious sometimes," she observed in her piece about the experience. "They'd all programmed their Lovegetys for the 'get romantically involved' mode."[16]

Eventually, the Lovegety fad fizzled out. Later sales figures revealed why: Kemp's experience was more common than Mikuriya's because a majority of the date-detecting devices were purchased by men.

Skip ahead to May 2004. The setting is CELab, a conference of tech industry executives and luminaries at MIT. Nathan Eagle, who is earning a PhD in wearable computers, is launching an experiment called Serendipity. Each participant at CELab picks up a Nokia 6600, Bluetooth-enabled, Symbian Series 60 phone. The phone is programmed to send icebreaker introductions, which Eagle describes in his thesis as "messages to two proximate individuals who don't know each other but probably should."[17]

The icebreakers are a bit like the Lovegety chimes in function, but are more complex in design. Each participant has created an online profile of herself describing what she does, what she's working on, and, in some cases, areas of expertise she's looking for.

A program called BlueDar (developed at the MIT Media Lab by Mat Laibowitz) is running silently in the background of the experiment. Every five minutes it scans the conference area for the locations of each phone so it knows who is standing next to whom. It does this by logging every device's media access control (MAC) address, a hexadecimal number unique to all Internet-enabled devices. Every gadget that collects or sends data over an Ethernet connection has one of these identifiers. While MAC addresses on desktop-bound machines have been around since 1980, the presence of MAC addresses on phones is, in 2004, an extremely recent phenomenon. Recent and significant.

"That means that my phone can recognize the fact that my laptop is within five meters [sixteen feet]. And the realization that everyone's carrying around these devices that are essentially broadcasting a unique ID means that suddenly you can do social and proximity-based applications," Eagle told me.

Because each of Eagle's devices has a particular signifier, and all are connected to a larger network, a lot more information can be conveyed about the person to whom each device is connected. Lovegety, remember, provided only three pieces of data: proximity within fifteen feet, gender, and interest in chatting, karaoke, or something more.

Serendipity features in-depth profiles of users and a better matching algorithm. Eagle used a Gaussian mixture model to detect proximity patterns between users and then correlated these patterns with relationship types. The result theoretically should have been better introductions between more like-minded people.

In his write-up of the experiment, Eagle says this was largely the case. A group of VIPs from one very large tech company were delighted to be introduced to a bunch of VIPs from the same company whom they had never before met. But the prettiest-girl-in-the-room syndrome experienced by Charlotte Kemp—with the Lovegety system directing multiple unsuitable matches to her—popped up again in Eagle's experiment. This time the pretty girl was none other than Nicholas Negroponte, who was forced to make a lot of small talk with a Microsoft executive he, in the words of Eagle, "didn't want to talk to." (Today, Negroponte doesn't remember the experience, but doesn't doubt that it happened precisely as Eagle describes.)

When Eagle deployed the system across the broader MIT campus, he discovered much the same occurrence. Of the one hundred people who participated in the experiment over the course of nine months, women were more cautious and concerned about privacy than were their male counterparts. The one group that most appreciated the experiment was the students from MIT's Sloan School of

Management, arguably the most stodgy and least technological population of the MIT student community. Sloan students were enthusiastic for the opportunity to better network with their peers across other departments. The problem, writes Eagle in his thesis, the other MITers "weren't as excited about getting introduced to Sloan students."[18]

Moving the activity of matching from the desktop PC to the real world was supposed to solve the prettiest-girl-in-the-room syndrome. Instead, it spread the problem, contaminating reality.

At the time of Eagle's experiment, large firms were projecting that 80 percent of new mobile phones sold would have Bluetooth capability within two years. "If that prediction holds true, applications like Serendipity would have the potential to transform dramatically the ways in which people meet and connect with each other," he wrote. "As technologies converge, new mobile phones can identify each other with Bluetooth and can re-create the functionality of the Lovegety by leveraging the information already stored in existing online profiles."[19]

The prediction did hold true. The launch of the iPhone and, later, the Google Android system have led to a flood of proximity-based social networking apps, as covered in chapter 2. In terms of matchmaking, all of these share the same fundamental flaw.

Consider this: one of OKCupid's key accomplishments was their mobile app which, after the gay male dating app Grindr, was one of the first dating apps to boast a location-broadcasting feature. Walk into a room, open the app, and—depending on how the other users around have configured their user settings—you get a window into the OKCupid users around you, their likes, preferences, histories, et cetera. One could easily imagine the app becoming the most downloaded program for such situational-awareness devices as the Google Glass headset. Picture yourself putting on a pair of goggles, glancing around a bar packed with singles, and seeing each person in light of how they answered questions about their previous sexual partners. The OKCupid mobile app represents the future of the way we'll interact with proximity-based social networks.

At least that's the way mobile dating apps with a GPS feature were supposed to work. In the last few years, reporters such as Lauren Silverman have observed what Charlotte Kemp discovered years ago: that with the exception of smart apps like Tinder, most geo-location-based dating apps are, like dating sites, mostly used by men. Women comprise just 36 percent of dating app users. It turns out women don't relish the thought of all their intimate secrets, along with their current location, rendered visible to any single man who happens to join an online dating site.

Yagan, however, is still convinced that mobile phone data is going to make online dating more fun and more effective. But finding you a date *at this moment* will be less important than improving your overall chances of finding a good date *in the future.*

"Dating is one of the categories of consumer Web products where the end goal is an in-person interaction. It comes into play in a way it wouldn't for your Twitter mobile app. We can leverage that in a more transformative way than other categories. Your dating app will go on your dates with you," says Yagan. The data that Yagan is seeking would look a lot like the customer feedback that eBay uses to rate transactions and sellers. In the case of Match .com, the product in question would be a date with someone from the site.

"Right now," Yagan continues, "because we don't know when dates take place, the algorithms optimize for on-site communication. It's hard for us to tell if a conversation you're having on the site is going to lead to a date or is just superficial chat. And because we don't know if you went on the date, we can't ask you for feedback on if the date went well."

Match.com's research suggests that people are relying a lot less on bars to meet people, so the service has gone into the business of hosting and sponsoring its own events, from happy hours to wine tastings to cooking—events that, according to Yagan, have attracted hundreds of thousands of people around the world.[20] "That's a service level that OKCupid doesn't offer but that Match does." Every day, people go to bars to meet other people. For these folks, a sushi-rolling class doesn't seem to be a better opportunity to meet your soul mate than a boozy happy hour, but because it offers a better opportunity for actual collaboration around a project, it probably is.

Bottom line, these services still suffer from the prettiest-girl-in-the-room syndrome but they've put researchers closer to curing it.

Introductions are important, but facilitating meet-cutes between people is only a first step in solving the deeper mystery of collaboration dynamics. Partnership, romantic and otherwise, is built on patterns of communication: What happened after who said what? How did A respond to B? How quickly? What was the tone? What does that mean? Our brain evolved expressly to help us navigate this highly chaotic give-and-take. In fact, research suggests that we evolved the ability to detect certain types of lies almost at the same time that we learned to vocally express complex ideas.[21] A truly quantitative approach to love requires a system that remembers *every* useful signal that's traded back and forth.

Sandy Pentland's Honest Signals

Alex "Sandy" Pentland, the head of the Human Dynamics Lab at MIT, has spent thousands of hours recording human interactions. Doing this in a lab setting would have been cost exorbitant and yielded results that were biased by the artificiality of the lab environment. So in 2002 Pentland made the world into his lab. He put together an infrared (IR) transceiver, an 8-kHz microphone, two accelerometers (the sensors in an iPhone that measures movement), a 256-MB flash drive, and four AAA batteries on a sensor board, encased in plastic. The first sociometer, as he calls these devices, was a bulky shoulder mount that looked like something a futuristic soldier would wear in a 1980s science-fiction movie. But it worked to record key aspects of human interaction, including pitch, volume, acceleration of speech, and use of hand gestures.

Pentland wasn't going to all this trouble to record the content of the conversations. There was no shortage of material that had been recorded for its content. Instead, he was out to record the signals we deliver and process unconsciously. He calls these signals "honest." You could also call them predictive.

In his 2008 book, *Honest Signals: How They Shape Our World*, he lays out four signaling types that make up the basis of human interaction: influence, "measured by the extent to which one person causes the other person's pattern of speaking to match their own pattern"; mimicry, which he calls reflexive copying of nods, smiles, other gestures; and activity, which includes a lot of hand usage, speaking quickly then slowly, softly then loudly, high and then low.[22]

These signals, working together like elements of a costume, serve to establish *character* in conversation in a way that's hard to consciously notice but is nonetheless deterministic. A lot of activity indicates sudden excitement, which further suggests that the actor is "open to being influenced." You're not all that confident in what you're saying but you're certainly interested in what the other person may say. The last signaling type, consistency, refers to evenness in tone. It's marked by focused attention and low movement. If you have a lot of it, this communicates a message of confidence and resolution.

How you and your conversational partner send or receive these signals indicates how interested you are in what the other person is saying, how open you are to her ideas or concerns, how certain you are about what's coming out of *your* mouth, and your level of interest in taking the conversation to a new place. In short, these signals tell your conversational partner what role you intend to play in the interaction.

Are you displaying focused attention, mimicry, and agreement? You're signaling that you will take the "teaming" role. Are you speaking first, modulating your voice little, and displaying muted activity? You're taking a "leading" position. Are you bouncing around in conversation, talking with your hands, fidgeting, and throwing in personal asides? This is the "exploring" role and accepting it means you're open to what the other person has to say. The final role, "active listening," works about as you would imagine that it would. None of these character parts is better or worse than any other. They all serve a precise function and we slip into and out of them at different times; it depends on whom we're talking to and the subject at hand.

With just thirty seconds of sociometer data, Pentland can predict what roles two people will take in a given interaction. Your brain takes about the same amount of time to unconsciously make the same prediction when you chat with someone. But unlike your brain, once Pentland has predicted the role, his model can anticipate what's going to happen, specifically who is going to get her way. And the model can do so with as high as 95 percent accuracy. Bosses who took the leadership role during salary negotiations with underlings were far more likely to walk away from the table with the better end of the deal. Pentland writes that this was a "better predictor of negotiated salary than any other factor."[23] Poker players who engage in active listening by showing their openness to being influenced and suppressing excitement are probably holding a much better hand than they're letting on. This is the amateur's poker face. It's largely the same around the world and arises from the attempt to suppress the fight-or-flight instinct that people experience when engaging in a high-stakes behavior, an adaptation that was paramount to our survival in the wild. When amateur players have a bad hand they act normally but when they have a particularly good hand they go wooden. "People are not very good at judging how much nervous energy they're showing, so they say to themselves, 'To avoid looking excited, I'll look like a dead fish,' " Pentland explained to me.

When a poker player tries to suppress the functioning of the entire automatic nervous system, she's using a very small portion of her brain to override the functioning of a very big part of the brain. A lot of energy gets used up. To compensate, the brain shuts down other actions, like fidgeting. The result is a rather unnatural stiffness that's largely absent when there are no strong stimuli to inspire the fight-or-flight response, as when you're playing your cards straight. Pentland observed that silence and stiffness were correlated with bluffing 68 and 67 percent of the time, respectively.[24]

He became so good at recognizing this signal that today he can effectively outperform virtually any semipro poker player. But he has no interest in that. He hates Las Vegas. He himself displays very few of the signals of a hard-nosed negotiator. Sandy Pentland is a terrible bluffer.

Pentland's office is on the fourth floor of the MIT Media Lab. When you meet him in person, after reading his book, you notice that his voice rises and falls easily. He can grow quite animated when he's discussing some aspect of his work about which he's particularly passionate. For someone who knows more about the value of adopting particular affectations than anyone else on the planet, he seems incapable of faking who he is. That's the entire point, according to Pentland. Our unconscious signals are our honest signals and a better indication of what we are really thinking, what we're really like, than anything we say. If Pentland, the man who discovered these signals, can't easily hide them, it seems there is little hope for the rest of us.

"What's cool is that now we can measure this stuff. And it's the same pattern we see in bees, in small human groups, and in large networks. It's a different way of thinking about things. Everybody else wants to look at the content of ideas. I just want to look at the flow of information. If the flow is good, good ideas will surface."

Once you learn to recognize honest signals, it's impossible not to see their impact everywhere. A great example is the 2012 U.S. presidential race. In the first debate Mitt Romney stuck with an even tone. His voice did not ascend or descend in either volume or pitch very often and he spoke with little hesitation even when he was going on the attack. Barack Obama, conversely, began many statements with long pauses underscored by the rasping sigh that has become fodder for mimics, thus displaying greater variance in pitch and voice than his opponent. Later independent analysis shows that the president made far fewer inaccurate statements than Mitt Romney did in that debate, but that's not how the public decided the winner. In surveys conducted immediately after the contest, people described Barack Obama as "disengaged." In the end this made it appear to the television audience as though Mitt Romney was in expert command of his material, especially when he was attacking the president. The president meanwhile looked insecure and unstudied even when defending his own record, subject matter about which he should have been deeply familiar.

Pentland admits a certain cynicism about the political process. He doesn't believe what candidates say in these sorts of exchanges but that doesn't mean the debates aren't good viewing if you know what to look for. "If you watch them this way, watch the meta-signals, you can tell what they get excited about. There's a little extra in there or a lack of something," he told me.

Pentland's work also shows that honest signaling is more important to romantic relationships than profile matching.

To demonstrate this he and one of his research partners, Anmol Madan, attended a local Boston speed-dating event and asked the participants, all between ages twenty-one and forty-five, to wear the sociometers (a somewhat later version that was smaller and less intrusive). They observed sixty five-minute sessions. The participants themselves were instructed not to exchange any numbers directly. Rather, at the end of the evening, every participant was told to pick which sessions they thought went well and then numbers could be released. Pentland and Madan found that by paying attention to social signals between the couples they could predict, with 73 percent accuracy, which women were going to release their numbers. (The baseline, flat-out guessing, will get you to 20 percent accuracy.) Mimicry was the most obvious giveaway. A woman who in some way modeled her conversational partner's speech or, in particular, his movements was the most likely to release her phone number. A slightly less important tell was variation in emphasis, speaking fast and slow, high and low, and revealing an openness to influence. Together these two modes of signaling were a "display of the exploring role."

Here's what was most significant about Pentland's study: the response of the male speed-daters could *also* be predicted on the basis of female signaled interest. While the male signaling had relatively little effect on the women, female signaling had a big effect on the men.

Pentland had expected that every male would release his number to every female he had a session with because everyone knows men make themselves more sexually available to everyone in order to maximize the chances for sex. Instead, what Pentland found was that the men were more likely to give their number to a female who had *also* elected to give her number to them.

Here is where the prettiest-girl-in-the-room syndrome met its match. The guys came in with their Lovegetys set to 10 but *left* with an understanding of how each interaction with each conversational partner unfolded, and that changed each guy's own level of interest. Every male liked the women who liked him back and lost interest in the women who seemed to be shooting him down.[25]

Anthropologist and bestselling author Helen Fisher, Match.com's chief scientist, is also interested in capturing unconscious signals that explain personality in a way that's a bit beyond what the average online dater even understands about herself or puts on a form. Fisher is leading a research effort to pin "dimensions of temperament" to individuals based on biology. In essence she's proposing a unified theory for personality. And because dating sites sell matches based on personality, it would be a big competitive edge for Match.com if her theory proves valid. She's looking to show that your dating personality is a function of your individual brain anatomy. She says that individual brain chemistry "could be useful in defining some of the primary biological structures of personality."

Her central argument is this: our personalities are governed by how different neurotransmitters interact. Dopamine-led people are, in Fisher's words, "curious and energetic." This category is for eccentrics, people who love to travel, can't stop asking questions, display a certain degree of impetuousness, and are eager to explore the world. It's the category in which she places herself. If you're more cautious and inclined to follow the rules, you're serotonin led.

An analytical and tough-minded arguer? Your temperament dimension is testosterone based. Finally there's the estrogen-based personality characterized by empathy and attentiveness. The estrogen profile is the quintessential nice guy or nice girl.

Fisher is not the first researcher to propose a matrix to pigeonhole people into different personality types. Ernest Tupes and Raymond Christal's five-factor model, which scores the personality on the basis of openness, conscientiousness, extroversion, agreeableness, and neuroticism is perhaps the most widely known. The fundamental flaw with these personality matrices is they aren't based on biology, says Fisher. They start with an assumption rather than a hypothesis that could be falsified with real evidence from fMRI data. "Make the questionnaire based on what you know from biology, and then go back to the biology to study whether, in fact, what you say you're studying, you actually are studying," she says. As she explains it, "The bottom line is, I am not studying your culture. I'm studying your biology. I'm only studying traits that I know have a biological basis."

To demonstrate the validity of her theory, Fisher and her colleagues conducted fMRI scans on a wide assortment of people and have also designed a questionnaire (she refers to this as the Fisher Temperament Inventory), which seeks to measure "the degree to which one expresses aspects of these four trait constellations."

She's also played a key role in a major survey effort at Match.com called Singles in America, which is America's largest survey thus far on dating habits. It's a first-of-its-kind big data study on dating, featuring responses from more than 107 million people who took the survey on Chemistry.com (also part of IAC's portfolio of dating sites) and Match.com. Among the findings from the survey released in the most recent study:

- A friends-with-benefits arrangement is becoming more mainstream every year. Some 44 percent of surveyed adults admitted to having been in a friends-with-benefits relationship in 2012, up from 20 percent in 2011.

- Your friend with benefits should be monogamous. More women than men say they would prefer to be in an exclusive relationship before going to bed with a partner (37 percent of single women in 2012, 31 percent in 2011, and 25 percent in 2010).
- Perhaps the best finding for Match.com's bottom line: the surveys show that more people are looking for love on the Internet first and other places second. Far more people meet online today than they do in bars. Nearly 30 percent of the singles surveyed had dated someone that they met online and 20 percent had met their most recent partner through a Web site. Only 7 percent said they had dated someone they met at a bar.[26]

"What big data can do for you is find patterns in behavior and personality that you would never find in a small sample—never find it because there's too much noise," says Fisher. "The vast majority of personality questionnaires, and all kinds of questionnaires, are based on the college population, for heaven's sake."

Like Yagan, Fisher is also interested in using data from mobile apps to take profile matching to an entirely new level. But she's looking for something rather more ambitious than a simple hookup app. What she wants is "the kind of app that uses what we know from evolutionary psychology, body language, linguistic studies, that really sums up a deeper understanding of who you're talking to . . . I mean, we spend our lives trying to size up the people around us constantly. And I think we're missing an awful lot of data, unless we start using apps that really do talk about word usage, body language."

This is what makes Pentland's work with the sociometer so potentially valuable to the future of dating and marriage. Because the sociometer provides a means to *continuously* see how your personality is affecting the person closest to you, it can provide a telemetric monitor for relationship health. After the experiment concluded, Anmol Madan actually created a program called a Jerk-O-Meter, which ran on a Zaurus VOIP phone.[27]

The program worked just as you might imagine. When the owner's attention or interest in the phone conversation began to flag, the program would send out such helpful notifications as "stop being a jerk." When Madan revealed the app, it received coverage from CNN, the *New York Times*, and other major media outlets, but it never emerged as a viable commercial product. Like Leonardo da Vinci's helicopter or Friendster, the program paid the price for being ahead of its time. The Zaurus was not a popular device commercially and the iPhone was still in its infancy at the time of development. Moreover, the idea of self-monitoring personal conversations for "degree of jerk" probably struck consumers as bizarre. People have been having conversations since, well, before there were people. Why should we now use a device to help us do the most natural thing in the world?

Today, Madan has his hands full with a new start-up, Ginger .io, which is applying telemetric signaling and predictive analytics to health. This arguably is a more noble cause than matchmaker tech. And the market for relationship software to help couples digitally analyze how they speak to each other doesn't really exist. But the utility of telemetric communication analysis is being proven. The testing ground just happens to be someplace other than love.

There's a reason people sometimes claim to feel married to their job. Marriage takes work, yes, but our work life has a lot in common with a long-term romantic relationship. Collaboration styles have a huge influence on outcome and performance. But there's a key difference: poor collaboration between two people in a workplace can hurt an entire organization, resulting in lost revenue or worse. This is why its employers are leading the way in developing techniques to actually collect real-time relationship data.

In the big data present, the honest signals that occur between people, the inaudible notes that make up the tone and character of our interaction beyond what is literally said, are mostly lost. In the naked future that ability spreads to more people and more couples. Suddenly, a lot of people can become much smarter about what effect their words and actions will have on the person they're with. This future is visible today in the way that a few ambitious organizations and companies are measuring collaboration dynamics.

How Office Relations Lead to Nuclear Meltdowns

In 2012, Cindy Caldwell, Christopher Larmey, and statistician Brett Matzke of the Pacific Northwest National Laboratory (PNNL) set out to try to predict where workplace accidents were going to occur around the lab and which teams of employees (or work groups) would be involved. An accident at PNNL is a bit more serious than a stubbed toe or a sprained wrist. The lab, which does work for the U.S. departments of Defense and Energy, is involved in cutting-edge research on nuclear fission, natural gas development, and weapons research. Employees handle volatile, radioactive, poisonous, and highly classified material on a daily basis. A bad day at the lab is a bad day indeed.

Matzke plotted all the accidents that had occurred in the lab during the previous year. He and his coresearchers had to consider all sorts of mishaps, from the ones involving explosive material to more mundane types involving vehicles, falling from ladders, or just misfiling paperwork. They wanted to see if there existed some common feature among them that predicted their occurrence.

They discovered that employees who indicated (via survey) that their relationship with their supervisor was strained, who felt they weren't well listened to, that their concerns weren't shared, and, as a result, weren't *engaged* in their workplace were much more likely to have an accident.

When Caldwell and her fellow researchers added together the scores for (A) whether the group worked with hazardous materials (note: they found that a work group will have 1.9 accidents a year just because they're exposed to hazards); (B) worker engagement; and (C) past operational experience (defined as previous incidents, sick days taken, staff performance, hire and attrition rate for a group), they were able to almost perfectly predict the *number* of incidents that each work group would experience that year.[28] Considering that the work groups contained an average of just sixteen people, it's a short leap from figuring out the weak link work groups to the particular weak link workers in the groups.

That number becomes far more useful if you can also predict *when* that accident will take place. When I asked Caldwell if there might be some way to do that, she acknowledged that constant telemetric monitoring of employee engagement would provide more actionable data than a once-a-year survey.

A couple of years ago a California company, e22 Alloy, began marketing a software-as-a-service (SaaS) application (called Alloy). The company was one of very few start-ups that could analyze the "continuous, objective data of the online activities of the workforce." The distinction between continuous data collection and spying is a subtle but important one. Monitoring employee behavior without that employee's knowledge and using collected information to punish employees can indeed be called spying. Forcing employees to submit to having all their computer activity watched is not spying if you tell them you're doing it but that sort of petty office-tyrant behavior isn't going to be good for morale and probably won't be much of a productivity boost, either.

Company founder Josh Gold was very sensitive about the spying applications of his product. In his presentation at Strata 2012, he recommended that employers not use the program without the explicit permission of their staff, and that employees should be able to suspend tracking whenever they choose.

Used properly, this sort of app could provide supervisors with "advance warning" that a big project is headed off the tracks. For instance, if you're a manager and one of your work teams starts communicating a lot more, but tangible work product decreases, that's a warning sign, as is "changing activity patterns," which could take the form of a lot more e-mails suddenly shooting back and forth at the end of the day and/or a lot of profile updating on LinkedIn.[29]

Back to love. In the same way that tone, timing, and particular changes in interoffice chatter can help predict project failure, communication changes between spouses can be indicative of buried problems. Marriage and work really do share a lot in common.

Here's a case in point. My wife and I both work from home and are copartners in a little joint hobby we call "not letting the house become an apocalyptic hellscape." Succeeding in this endeavor requires a certain amount of vacuuming, laundry folding, moving of trash and recycle bins to the curb, returning them to the side of the house, and dishwashing. It demands effort, management, and communication. In fact, it's very much like a regular job, and each of us has one (or more) of those as well.

More important, neither of us can agree on the definition of "hellscape." I fall more on the literal side. In fact, if I'm deep in a project, I may not notice that an interdimensional portal has opened over the cat-litter box until Yog-Sothoth the Outer God wraps his cold tentacle around my neck. My wife, meanwhile, knows by sight how many days it's been since someone ran the vacuum.

In our interactions, we'll fall into the role types outlined by Pentland's research. "Do you think we should run the vacuum?" she will ask, taking the explorer role, which my wife uses to talk to me about vacations to faraway places, contemporary issues, and our mutual acquaintances. I do not want to explore the issue of running the vacuum. I will respond that I am too busy and prattle off all the items on my to-do list I feel are more relevant. In doing so, I will speak calmly and evenly, taking the leadership position. My to-do list is something I can speak on with authority. I will win this exchange, but in doing so, I will lose. My wife will run the vacuum but not feel good about it. She has her own to-do list that is as long as mine, but she can make time to play "not letting the house become an apocalyptic hellscape" twice as hard. I detect her feelings and reflect them, rationalizing my own feelings of resentment.

People in long-term relationships approach exchanges with residual notions and emotions from the last exchange. Over time this can erode a person's ability to objectively perceive what's fair or logical in terms of the division of household labor, expenses, goals, and so on. Communication telemetry could fix this problem. Now imagine the workforce telemetry solution described above in place at home.

If I'm deeply involved in a project, running up on a deadline, then my communication patterns, my Internet usage, my verbal exchanges with my wife will indicate this just as it does when a workplace manager is floundering on a project. In those instances where I truly am too busy, I won't have to tell my wife I don't see the need to vacuum; she will actually be able to verify it herself. I, likewise, could use data from her communication patterns to reach a better understanding of her current stress level, which would speak to engagement with the house. *I don't ever have to see the house in the same way that she does to remember to run the vacuum.* All I have to see is her current stress level and then, without questioning, bring out the Dyson.

This isn't a perfect solution to the problems that arise in long-term cohabitation, but it does strike at one of the biggest unspoken problems in modern marriage. After a certain period, we expect the person we are with to be able to anticipate our moods. This expectation is not born of any rational or objective understanding of the way we communicate but of simple exhaustion. We become tired of explaining ourselves. The only solution is to develop the capacity to say more without exerting more effort, and telemetry can help with that.

In the years ahead, if more managers begin to experiment with telemetric solutions and if those experiments prove to be successful, we may become more accustomed to the idea of digging deeper into the secret signals inside our private communication. Not everyone will want to monitor their relationship signals for warning signs but for those who do, the process will become easier and cheaper. More of our talking, chatting, and signaling is taking place online and will be retrievable later at lower cost.

The last predictor of relationship longevity, according to Finkel, could be called the stress test. The late psychologist Reuben Hill first proposed it in 1949 after surveying couples who had been separated during World War II. What he found was that how a couple deals with unexpected emergencies—the loss of a child, critical injury, a sudden drop in wealth—portends strongly for the future of that couple. In much the same way reactions to stressful events predict future health states for individuals, couples who were able to survive high stress events grew more stable and became less likely to break up.[30]

Stress is the fiery crucible in which truly stable marriages are formed. We've long known this, but today the effects of stress on husbands and wives—or potential husbands and wives—can be modeled or run through a simulation, in the same way we simulate hurricanes, floods, and credit defaults to test the resilience of infrastructure and institutions. Some of the most promising work in treating post-traumatic stress disorder (PTSD) right now is based on running simulations of the traumatic events. There's no reason why a couple who was really serious about forging the strongest long-term relationship possible couldn't run simulations or game traumatic events beforehand to see how such an event would impact their relationship. If current experiments treating PTSD with simulations continue to prove effective, marriage counselors could recommend traumatic event role-playing as a means to better ensure relationship health.

Currently, scenario testing traumatic events is not an action that people consider when planning a future with someone. There just never seems to be a good time to tell the person that you are with, "Before we take this any further, let's do a few virtual reality natural disaster simulations!" In other words, there is no science or data yet on the effects of stress testing on marital relationship longevity. That's just not the way we think of love, but we *know* that stress tests and simulations in business and engineering are effective in finding problems before those problems blow up. No one would get on a bridge that was marked with a sign reading THIS BRIDGE HAS NEVER BEEN LOAD-TESTED. Yet we carry on for years in relationships to which we've never applied any sort of objective strength test outside a *Cosmo* quiz. When we learn to approach personal relationship decisions with the same seriousness that we collectively approach issues of public safety, then all of us will experience fewer relationship disasters. The idea might seem far-fetched but a few years ago so did the notion that a majority of singles would turn to the Internet to find love before heading out to a local bar. As stress tests and virtual reality simulations prove their utility in other areas of life, we will eventually get around to applying them to love and then the last component of the formula will be in place. We can finally create our soul-mate predictor app.

The Love Machine

So what is the future of love? We know that personality profiles can help you predict who will or won't be a great date; sociometrics can tell you how well a date will go. A data set of sociometric scores going back for years will reveal how your personality, and that of someone else, might interact over a period of years. Trauma simulation can even give you a sense of how your marriage will weather life's big storms. An ensemble of these scores won't tell you if you're in love, but you can predict arguments and resolve them in advance. You can get a window into the future of your relationship with someone. You can find a mate who is indeed scientifically suited to you. More important, you can use science to make your relationship stronger.

The first step toward a better science of dating is getting customers to give up more data about themselves and how they date, beyond simple information about the sort of person whom they're interested in.

What Yagan wants is a lot more data from his users, not just information on how they answer questions about what they're looking for in a relationship but also Amazon reviews that provide a sense of why some people find some products are superior to others, Facebook and Foursquare information about comings and goings, Fitbit data to measure the beating heart. These signals, formerly inaudible but now detectable online, make up what Yagan calls "true identity" and leveraging true identity "broaches a line that no site has managed to cross."

Sandy Pentland reached a similar realization early on in his work on sociometric data; that "by adjusting for personal characteristics and the history of previous interactions, we can dramatically increase our ability to predict people's behavior."[31] With enough data a naked future emerges, a profile that is more living and thus predictive than any survey questionnaire because it is assembled from action, because it changes, as do you and I.

But would we dare call this love?

Perhaps we need to change our definition of what love is. We tend to view it as something we own and thus can lose, something we want and are entitled to, and something we lend in the hope of getting back. Perhaps love is more fluid, less connected to who we are and more firmly attached to what we do. Love is more than dopamine (at first) and oxytocin (later). It's a decision that, if we are lucky, we are called upon to make over and over again. We make hundreds of decisions in our relationships every day. If we could develop the ability to pick up just a few more of the signals that the person we love sends out continuously, then that decision making would improve. Love becomes less work.

Though the Brahmins understood little about the makeup of the universe compared with what we know today, they understood that idea well enough.

After describing why Vedic astrology is an expert practice worthy of admiration, Paramahansa Yogananda, in his *Autobiography of a Yogi*, effectively devalues the entire endeavor to predict the future and launches an eloquent defense of free will: "The message boldly blazoned across the heavens at the moment of birth is not meant to emphasize fate, the result of past good and evil, but to arouse man's will to escape from his universal thralldom. What he has done, he can undo. None other than himself was the instigator of the causes of whatever effects are now prevalent in his life. He can overcome any limitation, because he created it by his own actions in the first place."[32]

CHAPTER 9
Crime Prediction: The Where and the When

WHEN crack first got to Pittsburgh, Pennsylvania, in the late 1980s, police attacked the problem the only way they knew how: busting dealers who were working out in the open, performing sting operations, and planting patrols on blocks where they had disrupted drug traffic before. Getting a dealer off a particular block was a big victory. If you're a Pittsburgh crack pusher, you can't just send a letter to your clients with your new address. And naturally, as anyone who has ever seen *The Wire* knows, whenever a dealer is forced to relocate to a new block he runs the risk of encroaching on territory that belongs to another dealer, which can lead to . . . disagreement. The police understood that clearing and holding blocks were crucial to slowing the spread of crack but they didn't have the resources to clear and hold *every* neighborhood. Some dealers were going to relocate. Finding a new market that isn't yet occupied buys a drug dealer a lot of time to set up. If the police could anticipate which neighborhoods were the most conducive to drug dealing and why, theoretically they could predict where the dealers were going to go set up.

How to figure this out? The most well-established approach to predicting which neighborhoods were going to experience an uptick in crime was called broken-windows theory. In 1982 researchers James Q. Wilson and George L. Kelling observed a correlation between neighborhood dereliction, vandalism, vacancy, little lifestyle crimes like prostitution and panhandling, and broad crime increases. Neighborhood dereliction took the form of broken windows. To this day, the theory remains the basis for the zero-tolerance police efforts in places such as New York under the Giuliani administration and, to a lesser extent, Baltimore under former mayor Martin O'Malley. While it offered an effective if controversial approach for mayors looking to appear tough on crime, it was a lousy tool for predicting what *sorts* of crimes were going to take place and where and when they were likely to occur. Exactly how many windows have to be broken in your neighborhood before a crack dealer sets up on your corner?[1]

Enter Andreas Olligschlaeger, a systems researcher and public policy scholar at Carnegie Mellon University in downtown Pittsburgh. He knew that certain variables make a neighborhood attractive for a drug dealer looking for new turf. One factor is the presence of commercially zoned space. Passersby are much less likely to call police on potential drug dealing near a factory or warehouse than by their own homes (and there are also fewer people around at night). The next factor was seasonality. Drug dealing is mostly an outdoor activity and tends, like cherry trees, to blossom in the spring and flourish in the summer.

Olligschlaeger also knew that the number of 911 calls related to weapons (shots fired), robberies, and assaults provided an indication of an emerging drug-dealing neighborhood. All these elements were clues as to where the pushers were going to go. The question became how to weight those variables. Was the presence of a potential competitor in one neighborhood more or less of a factor than a lot of residents hanging around? Exactly how big a role did seasonality play? And were the variables dependent or independent? Did an assault in one neighborhood affect the attractiveness of another neighborhood as a new drug-dealing spot or did it not matter? At the time, Pittsburgh had a computer system called DMAP that allowed for the tracking of crimes across geographical space. But a straight averaging of these factors would likely result in a model that treated all the variables too equally. It would overfit.

Classical statistics doesn't lend itself well to modeling chaotic interactions with lots of moving parts, but artificial neural networks, which were a relatively recent innovation in 1991, were showing some interesting promise in the field of high-energy physics. An artificial neural network (aka neural net) is a mathematical program modeled on the way neurons exchange signals in the human brain. One of the core features of a neural net is that the weighting of variables changes as the system processes the problem repeatedly. In the same way that a kid shooting free throws from the same spot eventually becomes a great free-throw shooter, or the novice artist who does a thousand different sketches of hands becomes a better artist, neural nets learn by applying a particular set of tools to a particular problem over and over again. Though he was specializing in public policy at the time, Olligschlaeger is also the sort of guy who reads physics journals, which turned out to be a good thing.

The movement of drug dealers around Pittsburgh had to be subject to mathematical laws just like the movement of particles. Olligschlaeger trained a neural net system on every 911 call related to assault, robbery, and weapons from 1990 to 1992, as well as six other variables (for a total of nine) and then ran fifteen thousand simulations. The system came up with a series of predictions for which 2,150-square-feet sections of the city would see an uptick in drug-related 911 calls. At the end of August 1992 Olligschlaeger made three maps: one showed the predictions made by the straight statistical model (regression forecasts), the other two were neural net based. The result was that the neural net model presented a clear 54 percent improvement over the straight averaging model.[2]

The models all performed differently, and none predicted the actual number of calls perfectly. But the straight statistical regression model overestimated the number of calls that occurred by a great deal so if the PD had used that model, they would have sent a lot of cops to quiet neighborhoods in anticipation of calls that would not come. That means they wouldn't be covering the problem areas as well. The neural net, conversely, missed the relatively few calls that occurred in the southwestern portion of the city. Yet compared with the statistical regression model, it was the model that was *least* likely to send police to a place where there was definitely not going to be any action. It did a much better job predicting not only where crime would *not* be but also the number of drug-related calls that would occur in each map cell where they did happen. It provided far better value than straight guessing or even traditional statistical analysis. Unfortunately, that wasn't good enough to convince the city of Pittsburgh. They never adopted neural nets as a crime-fighting tool.[3]

You can't blame the city hall bureaucrats for not buying the neural net concept. The connection between the input (data) and the output (prediction) was too opaque. Even though the predictions themselves were good, the lack of transparency as to how the net reached its conclusion made the entire system unattractive from a policy standpoint. In his paper on the subject, Olligschlaeger himself admitted this: "One disadvantage of neural networks is that there currently are no tests of statistical significance for the estimated weight structures. However, if the main goal of a model is to provide good forecasts rather than to analyze relationships between dependent and independent variables, then this should not be an issue."

Though the use of neural nets did not become standard practice, Olligschlaeger's study represents a key evolutionary moment of what is today called predictive policing, the use of computational databases and statistics to identify emergent crime patterns and deploy cops preemptively.

Skip ahead to 1994, newly appointed New York City police commissioner William Bratton institutes what he called a "strategic re-engineering" of the city's police department. The use of up-to-the-minute data, citywide crime statistics, and crime mapping will go on to bring down the city crime rate by 37 percent in three years. Bratton's reengineering became another important victory for predictive policing, but not a decisive one because stats were only one portion of Bratton's overall strategy. Today, many scholars credit tougher zero-tolerance and stop-and-frisk policies, coupled with the use of crime mapping, for bringing down New York City's crime rate in the 1990s. These measures were not without controversy. New York's aggressive law enforcement strategies under Bratton led to complaints and charges of harassment and overly aggressive tactics, particularly the stop-and-frisk provision, which targeted mainly minority youth.[4]

The first unequivocal victory for predictive policing in practice occurred in 2003 in Richmond, Virginia. Criminologist Colleen McCue was using IBM's Statistical Package for the Social Sciences (SPSS) software as part of her research into crime patterns. She realized that incidents of random gunfire around New Year's Day in Richmond happened within a very specific time period, between 10 P.M. on New Year's Eve and 2 A.M. on New Year's Day. And these incidents occurred in very particular neighborhoods and under unique conditions. With these variables she built a model to show that on New Year's Eve the department could dramatically cut down on gunfire complaints, nab a lot of illegal firearms, and do it all with far fewer officers than they had used for patrol the year before by placing police in the places where the gunfire was most likely to occur.[5]

Most police departments are run like regular businesses. Cops have precincts to report to and are scheduled in regular shifts. Cops go out on patrol to look for crimes in progress but most of the job is responding to complaints and calls that have come in. The idea of sticking a lot of cops in one spot, in one time window, in advance of something that *might* happen was pretty revolutionary in 2003, but the department followed her lead.

When the initiative dubbed Project Exile was concluded, gunfire complaints were down by half compared to the previous year, gun seizures were up 246 percent, and the department had saved $15,000 in New Year's overtime pay for officers. Complaints were down, guns came off the streets in droves, and more cops got the night off. It was a triple score.[6]

Both Project Exile and the neural nets showed that they could get results. Yet where Olligschlaeger found resistance from city officials in Pittsburgh, Richmond police were eager to embrace Project Exile. The reason why says a lot about the way city governments work. Because Exile didn't involve a neural net or any outrageously sophisticated modeling technique and was instead a straight statistical regression, the political decision makers could understand it. Neural nets are sometimes referred to as black box systems. It's extremely difficult to see exactly how they reach the conclusions that they reach. What was a fascinating system scientifically was unusable as a decision-making tool for a lawmaker or police representative, someone who had to be able to show how and why he arrived at a particular decision, almost regardless of whether the decision was right or wrong. Yes, Exile proved extremely effective when applied to the problem of random gunfire but the challenge of identifying emerging drug neighborhoods was rather more difficult and potentially of greater long-term significance.

Project Exile simply capitalized on better record keeping techniques. It worked on correlation. Using data to predict crime on the basis of cause was a much more important test. It would occur a few years later in Memphis.

The Red Dot of Crime

Over the last several decades, Memphis has followed the same path—straight down—of many formerly prosperous U.S. metro regions. Property values and college graduation rates are abysmal. Poverty is high. Throughout the early 2000s, Memphis was consistently ranked one of the top five worst U.S. cities for violent crime.

Between 2006 and 2010, in spite of all of the above, crime went down 31 percent.

The demographics of Memphis didn't change in that time. The approximately twenty-three hundred men and women on the police force at that time were the same sort you find in any town where there's too much to do and too few to do it. Here's what changed: the department began handling its information differently thanks to Dr. Richard Janikowski, an associate professor in the Department of Criminology and Criminal Justice at the University of Memphis.

Janikowski convinced local police head Larry Godwin to allow him to study the department's arrest records. But Janikowski wasn't looking for biographical sketches of the perpetrators; he was looking for marginalia, the circumstances behind each arrest, the *where* and *when* of crime.

The biggest single finding and by far the most controversial was that the rising crime rate was closely connected to Section 8 housing, federally subsidized housing for qualified individuals below a certain income level. When Janikowski and his wife, housing expert Phyllis Betts, took a crime hot-spot map and layered it on top of the map for Section 8 housing, the pattern was unmistakable. Hanna Rosin, in her 2008 *Atlantic* article on Janikowski, described it thusly: "On the merged map, dense violent-crime areas are shaded dark blue, and Section 8 addresses are represented by little red dots. All of the dark-blue areas are covered in little red dots, like bursts of gunfire. The rest of the city has almost no dots."[7]

When I asked Janikowski about it, he points out that the blue-area, red-dot analysis omitted some important data. "You know, the stuff didn't overlap perfectly. There were high levels of correlation with it. Section 8 housing was part of what you see there. But it was also just heavy levels—a big concentration of poverty. And that's a complex relationship that was occurring."

Today, we know with more certainty that the connection between Section 8 housing and rising crime is correlative, not causative. People who live in this housing are not more likely to commit crimes so much as they are more likely to move to low-rent neighborhoods where the probability of a crime rise is already high.[8]

This relationship between the likelihood of being a crime victim, being a crime assailant, and living in Section 8 was particularly complicated in Memphis, says Janikowski, where many traditional Section 8 units were in terrible shape and others were being torn down. "You've got lots of demolition that was occurring in what was the traditional inner city for various reasons. So you had movement there. You had a lot of movement of at-risk populations. And they all tended to cluster because, again, the price of housing."

Even if the chain of causation between housing vouchers and violent crime wasn't clear, the relationship was still a useful guide for predicting where crime was going to occur. Janikowski had to make this case.

He sat down at a local cafeteria with Memphis police director Larry Godwin, local district attorney Bill Gibbons, and representatives from the department's Organized Crime Unit. Janikowski was blunt. He told them that to better focus their efforts and get more value for their money, they had to go back over arrest records and take a better look at when and where crimes were occurring.[9]

Operation Blue CRUSH (Crime Reduction Utilizing Statistical History) was born. The system used IBM's SPSS program and mapping software from Esri to better capture and disseminate crime data. When the initial test in an area in East Memphis called Hickory Hill proved successful at bringing down crime at less expense, the department increased the number of police working the tourist areas around Beale Street after 11 P.M., then they focused on the relatively rough-and-tumble area that is today called Legends Park but that at the time was a seventy-year-old, soon-to-be-condemned housing development called Dixie Homes.

Blue CRUSH uses primarily rule-induction algorithms. In terms of complexity these lie somewhere between a neural net and a straight statistical regression. It's a learning program that comes up with its own rules for what different variables should weigh based on training data its programmers have exposed it to (this process of coming up with rules is the induction part). It's still a varying weight model but one with more traceable results.

The Memphis PD also looked at a lot more variables than the nine (or so) different factors that Olligschlaeger modeled. In addition to weather patterns, seasonality, and area demographics, they could also model lighting conditions with a particular focus on garages and alleys. They looked at when big local employers issued paychecks by time of the week, the month, and the year and what times of day people went to and left work.

The same location optimization techniques that companies such as Esri provide to retail chains to find the best neighborhood to place a new store are also useful in mapping relationships between crime, economics, and physical space. "We can not only just manage what is this dot on the map that we call 'burglary' or 'robbery,' but how does that dot on the map interact with the demographics of the area, home values or population trends," said Mike King, Esri's national law enforcement manager. "If you're in a predominantly blue-collar neighborhood that works at factories, what happens every other Friday when it's payroll time? Do we see increases in alcohol-related events? Do we see increases in domestic violence?"

Here's why the way these models work matters to the naked future: as we develop the capacity to monitor more of these signals and incorporate more variables, the statistical tools required to make use of them will become simpler and more transparent. (It's hard to conceive of practitioners today using a neural net, which is considered rather quaint.) As transparency increases, governmental decision makers will have an easier time accepting and supporting predictive policing programs. As more departments begin to use such programs, and share information about which variables and tools are most useful, these programs could get a lot better very quickly.

Changes in area economics have emerged as a useful signal for future crime predicting, but it's not a *clear* signal. If a sizable portion of the people in your neighborhood suddenly can't afford to pay their phone bills, or are facing vehicle repossession, that can be indicative of more potential criminals since clearly these people have fallen on rough times. But a sudden rise of neighborhood inequality is also an indicator since part of the neighborhood now *perceives* itself to be less well-off compared with its neighbors. Criminality, like envy, can be contagious.

In 2005 a military base reconstruction project left many residents of a particular San Antonio neighborhood suddenly a lot better-off than their neighbors. A big base realignment and closing program resulted in a bump in demand for a very particular type of contractor. Neighborhoods that had been fairly uniform economically were suddenly divided into haves and have-nots. Cornell researchers Matthew Freedman and Emily G. Owens showed that "because of the targeted nature of the spending program, an important effect of this program was to increase the criminal opportunities of the average San Antonian."[10]

But suddenly losing your job can also make you more likely to become a *victim* of crime. Janikowski found that when a group of women in Memphis who couldn't afford a landline were forced to make telephone calls from a pay phone on the side of a convenience store, their risk of suffering sexual assault increased.

One of the strongest indicators that crime in a given neighborhood is about to jump is foreclosures. Foreclosed homes invite burglars who ransack the residences for copper in wires and electrical equipment. Drug dealers seem to feel relatively comfortable working in a neighborhood the residents have been pushed out of by landlords and banks. Focusing on foreclosure clusters, and putting cops nearby as soon as the cluster appears, is broken-windows theory 2.0. Rather than react to neighborhood dereliction, it anticipates it.

Within the broader variable of seasonality, there's a lot of nuance. When the Memphis department focused a heavy police presence downtown during the end of summer, they were able to preempt a rash of burglaries and vehicle break-ins that would normally have been perpetrated by teenagers about to go back to school. In one week the PD dropped crime in that precinct by 37 percent compared with the previous year.[11]

The data collection and the analysis that made these predictive insights possible are accelerating and becoming cheaper. Mobile computing and the Internet of Things are allowing officers in the field to collect and disseminate incident data, and better access data from one another, much faster.

Today, police officers on the beat have the same rapidly evolving view of potential hot spots that headquarter dispatchers had a few years ago. Big command and control centers are moving away from situation rooms, where operators on headsets feed information to soldiers on the ground, and into a single console that patrol officers carry with them. The goal to make that information assimilation process work in a mobile environment is one of the key jobs of Mike King at Esri.

For instance, let's say you're a cop looking for a robbery suspect late at night. You know the general part of town the perpetrator is in but need more information to nail down a location. Let's say you have access to a big data set indicating tens of thousands of arrest records and you can query that database to learn the type of place that most suspects of this crime go after a robbery. It could be home, girlfriend's house, mom's house, bar, et cetera. Let's say you can also bring up a map to show you all of the closed-circuit television (CCTV) cameras in the area of the incident and even which stores are open late, where you might be able to find a witness who saw something. You can further ask a network of community members and other cops to mark on a digital map where they saw something that could be useful—an article of clothing left behind by the subject, a stolen item, a sighting of someone matching the subject's description. You now have not just one map but several that you combine to rapidly narrow down where to find your suspect and even obtain the evidence for conviction. Esri, working with police departments around the world, is putting that command center view on laptops and even phones. This is the challenge that occupies King: "How do I get that information into the officer's hands so that he can be there at that same time?" The NYPD, in partnership with Microsoft, has also been developing these sorts of capabilities for New York City beat cops through a program called "the dashboard."[12]

To understand how that contributes to a more naked future, you have to imagine that dashboard dissemination of capability continuing, eventually, on to consumers. In the same way computers used to be the size of rooms and were available for experts to use, then became objects people could access on desktops, and are now objects in our pockets, the dissemination of this type of big police data is going to follow the same path.

In the next ten years there's no technological, economic, or even legal reason why every individual with a smartphone shouldn't be able to download a live crime map showing both current and expected hot spots. Predictive policing won't just be something that happens around you, it will be a process that you participate in directly. Information will grow much more rich and meaningful when it's combined with other bits of local data and personal information. I asked Mike King if he saw this eventuality as likely. "It's happening today," he answered. "When I talk about Esri moving through mobile and other opportunities, that's the idea of getting this water to the end of the row."

Ever better and more timely reporting of crimes and incidents are key to the continual improvement in predictive policing. Increasingly, that reporting is being done by bots.

New York, Milwaukee, and nearly seventy other cities around the United States use a sensor network system called ShotSpotter, which uses acoustics to detect and pinpoint gunfire the moment the trigger is pulled. In California and along the eastern seaboard, cameras snap pictures of license plates as cars enter and leave specific areas of various cities. That's on top of a growing CCTV infrastructure, which, in 2011, comprised more than 45 million systems around the globe. Growing by 33 percent per year, the global CCTV market is forecast to become a $3.2 billion per year industry by 2016.[13] Not all of those cameras will be attached to buildings. In

2015 police departments around the country will begin testing aerial drones to establish a permanent eye in the sky in cities around the country, as authorized by H.R. 658, the FAA Modernization and Reform Act of 2012 signed into law by President Obama.[14] As a result of this bill, thirty thousand unmanned aerial vehicles will be crisscrossing America by 2020, according to Todd Humphreys of the University of Texas at Austin.[15]

There's a huge industry incentive to make it seem as though the growing web of cameras, microphones, sensors, and robot planes keeping watch over us is making us safer. Unfortunately, predictive policing won't automatically fix any of the long-term issues that plague our criminal justice system or change the way many cops interact with residents in poor neighborhoods. Zero-tolerance policies—of which predictive policing programs often serve as a component—are really effective at putting people behind bars. In a country with the highest prison population rate on the planet, that's like taking a machine that produces a terrible product, say, exploding strollers, and "improving" it not by changing the design of the strollers but by enabling it to produce many more exploding strollers far faster and more cheaply. Even in places where every criminal is truly a threat to public health (which is no place), pumped-up arrests will exacerbate prison overcrowding, recidivism, and so forth.[16]

But the most striking abuses of predictive policing programs and surveillance in general will likely soon emerge from China. China will surpass the United States as the world's largest market for surveillance equipment by 2014, according to a report from the Homeland Security Research Corporation (HSRC). The manufacturing hub of Guangdong Province, which is near Hong Kong, boasts a $1-million-camera-security system. China is today spending more on internal security than it is on defense, but many in the West, including NATO and the U.S. DOD, claim that Chinese military funding and "public safety" funding overlap a great deal.

Predictive policing in the wrong hands looks less like a boon to public safety and more like a totalitarian hammer. Some predictive policing tactics have already been used to stifle dissent and protest in the United States. In 2003 Miami police targeted and arrested several demonstrators prior to a major protest against the Free Trade Area of the Americas (FTAA) agreement. Today, police around the country routinely employ espionage tactics to predict and preempt spontaneous punk and dance shows (under the expansive and poorly written 2003 RAVE Act, sponsored by Joe Biden, which can be used to arrest concert promoters for the behavior of their patrons). If you're a police chief or mayor, preempting a protest is less risky than trying to disrupt one in progress, especially in an age where the kids you will be pepper spraying carry TV studios in their pockets.

This is bigger than busting garage punk shows, squashing Occupy marches before they take place, and shutting down raves before the speakers are even plugged in. It's bigger than the enforcement of vaguely worded local nuisance ordinances. The same tactics that can give police advance awareness of local protest events can also be used to predict civil demonstrations, marches, and clashes halfway around the world.

Acting locally is now visible globally.

Seeing the Riot Through the Trees

The date is June 30, 2012. Computer scientist Naren Ramakrishnan is in his Virginia Tech lab watching a map of the Americas on his computer screen. A band of hundreds of red dots hovers over Mexico City; another band is over the Brazil-Paraguay border. The dot cluster is ringed by concentric circles of yellow, green, and blue. It looks almost like a radiant heat map, as though the capital of Mexico and the Brazilian border town of Foz do Iguaçu are on fire, but they aren't—at least, not yet. These dots represent geo-tagged tweets containing the terms *"país," "trabajador," "trabaj,"* "president," and "protest." The controversial Enrique Peña Nieto is about to be officially elected the president of Mexico and the geotagged tweets represent a march taking form to protest his election.

In 2012 Nieto represented the return to power of the Partido Revolucionario Institucional (PRI). Despite the insurgent-sounding moniker, the PRI is very much the old-power party in Mexico, having governed the country for seventy-one years until 2000. It has long been associated with chronic corruption and even collusion with drug cartels. Nieto, a young, handsome, not conspicuously bright former governor of the state of México is seen by many as something of a figurehead for a murky, well-funded machine. Having met him I can attest that he can be very charming, smiles easily, and has a firm handshake. As a governor, he is best known for allowing a particularly brutal army assault on protestors in the city of San Salvador Atenco. The June 30 red-dot cluster over Mexico indicates a lit fuse around the topic of Nieto on Twitter.

At 11:15 P.M., on July 1, as soon as the election is called for the PRI, the student movement group Yo Soy 132 (I Am 132) will spring into action, challenging the results and accusing the PRI of fraud and voter suppression.[17] The next month will be marked by massive protests, marches, clashes with police, and arrests. This is the future that these red dots on Ramakrishnan's monitor foretell.

The cluster in Brazil relates to a sudden rise in the use of *"país,"* "protest," *"empres," "ciudad,"* and *"gobiern."* In a few days twenty-five hundred people will close the Friendship Bridge connecting the Brazilian city of Foz de Iguaçu to the Paraguayan Ciudad del Este, another episode in the impeachment drama of Paraguayan president Fernando Lugo.

As soon as clusters appear on Ramakrishnan's computer, the system automatically sends an alert to government analysts with the Intelligence Advanced Research Projects Activity (IARPA), which is funding Ramakrishnan through a program called Open Source Indicators (OSI). The program seeks to use available public data to model potential future events before they happen. Ramakrishnan and his team are one of several candidates competing for IARPA funds for further development. The different teams are evaluated monthly on the basis of what their predictions were, how much lead time the prediction provided, confidence in the prediction, and other factors.[18]

The OSI program is a descendant to the intelligence practice of analyzing "chatter," a method of surveillance that first emerged during the cold war. U.S. intelligence agents would listen in on the Soviet military communication network for clues about impending actions or troop movements. Most of this overheard talk was unremarkable but when the amount of chatter between missile silo personnel and military headquarters increased, this indicated that a big military exercise was about to get under way.[19] This analysis was a purely human endeavor and a fairly straightforward one, with one enemy, one network to watch, and one set of events to watch out for.

In the post-9/11 world, where—we are told—potential enemies are everywhere and threats are too numerous to mention, the IARPA considers any event related to "population-level changes in communication, consumption, and movement" worthy of predicting. That could include a commodity-price explosion; a civil war; a disease outbreak; the election of a fringe candidate to an allied nation's parliament; anything that could impact either U.S. interests, security, or both. The hope is that if such events can be seen in advance, their potential impact can then be calculated, different responses can be simulated, and decision makers can then select the best action.

What this means is that the amount of *potentially* useful data has grown to encompass a far greater number of signals. For U.S. intelligence personnel, Facebook, Twitter, and other social networks now serve the role that chatter served during the cold war. But as Ramakrishnan admits, Facebook probably is not where the next major national security threat is going to pop up. So intelligence actively monitors about twenty thousand blogs, RSS feeds, and other sources of information in the same way newsroom reporters constantly watch AP bulletins and listen to police scanners to catch late-breaking developments.

In looking for potential geopolitical hot spots, researchers also watch out for many of the broken-window signals that play a role in neighborhood predictive policing, but on a global scale. The number of cars in hospital parking lots in a major city can suggest an emerging health crisis, as can a sudden jump in school absences. Even brush clearing or road building can predict an event of geopolitical consequence.

Spend enough time on Google Maps and you can spot a war in the making.

Between January and April 2011, a group of Harvard researchers with the George Clooney–funded Satellite Sentinel Project (SSP) used publicly available satellite images to effectively predict that the Sudanese Armed Forces (SAF) were going to stage a military invasion of the disputed area of Abyei within the coming months. The giveaway wasn't tank or troop buildup on the border. Sudan began building wider, less flood-prone roads toward the target, the kind you would use to transport big equipment such as oil tankers. But there was no oil near where the SAF was working. "These roads indicated the intent to deploy armored units and other heavy vehicles south towards Abyei during the rainy season," SSP researchers wrote in their final report on the incident.[20] True to their prediction, the SAF began burning border villages in March and initiated a formal invasion on May 19 of that year.

Correctly forecasting a military invasion in Africa used to be the sort of thing only a superpower could do; now it's a semester project for Harvard students.

Much of this data is hiding in plain sight, in reports already written and filed. In 2012 a group of British researchers applied a statistical model to the diaries and dispatches of soldiers in Afghanistan, obtained through the WikiLeaks project. They created a formula to predict future violence levels based on how troops described their firefights in their diaries. The model correctly (though retroactively) predicted an uptick in violence in 2010.[21]

Simple news reports when observed on a massive scale can reveal information that isn't explicit in any single news item. As I originally wrote for the *Futurist* magazine, a researcher named Kalev Leetaru was able to retroactively pinpoint the location of Osama bin Laden within a 124-mile radius of Abbottabad, Pakistan, where the terrorist leader was eventually found. He found that almost half of the news reports mentioning Bin Laden included the words "Islamabad" and "Peshawar," two key cities in northern Pakistan. While only one news report mentioned Abbottabad (in the context of a terrorist player who had been captured there), Abbottabad is located easily within 124 miles of the two key cities. In a separate attempt to predict geopolitical events from news reports, Leetaru also used a thirty-year archive of global news put through a learning algorithm to detect "tone" in the news stories (the number of negatively charged words versus positively charged words) along fifteen hundred dimensions and ran the model on the Nautilus, a large, shared-memory supercomputer capable of running more than 8.2 trillion floating point operations per second. Leetaru's model also retroactively predicted the social uprisings in Egypt, Tunisia, and Libya.[22]

News reports, tweets, and media tone are *correlated* with violence. Predicting the actual *cause* of violence is more difficult. Yet researchers are making progress here as well. In Libya, Tunisia, and Egypt, the price of food, as measured by the food price index of the Food and Agriculture Organization of the United Nations (FAO), clearly plays a critical role in civil unrest. In 2008 an advance in this index of more than sixty base points easily preceded a number of low-intensity "food riots." Prices collapsed and then bounced back just before the 2011 Arab Spring events in Tunisia, Libya, and Egypt.[23]

If you're a humanitarian NGO, knowing where and when civil unrest is going to strike can help you position relief resources and staff in advance. If you're a company, you can pull your business interests out of a country where the shit's about to hit the fan. But to law enforcement, predicting the time and place of an event of significance is less important than knowing who will be involved.

Unlike predicting an invasion, piecing together a model of what a particular individual will do involves a lot more variables. Not only is it more challenging technically, it's also more costly. Researchers can't just run lab experiments on who will or won't commit a crime, so research has to take place in the real world. But experimentation here runs up against privacy concerns. In recent years researchers have found a clever way around these thorny issues by looking toward captive audiences, individuals in situations who have effectively relinquished any expectation of privacy.

CHAPTER 10
Crime: Predicting the Who

THE date is the Wednesday before Thanksgiving 2025. The setting is Dulles International Airport. Today is the busiest travel day of the year and the airport is crowded with parents dragging children dragging stuffed animals from gate to gate. But while there is no shortage of people in the airport, a single key feature distinguishes it from a similar setting as we would encounter it today. The people aren't standing in line. Nor are they attempting the difficult task of disrobing while moving through an X-ray machine. They aren't carrying their belts or shoes or being patted down by TSA agents. They're just walking to where they need to be or else waiting for their plane to board. There seems to be no security whatsoever.

The only apparent bottleneck in the flow of traffic occurs near the entrance to the departure gates, a spot where, just a few years ago, patrons would have encountered enormous detectors of metal, gas, and powder. Instead, visitors to the future Dulles are asked to walk up to a machine, stare directly into a lens, and answer a few questions.

The visitors approach the kiosk one after another, perform the task, and are moved quickly through . . . save one man, whose responses to the requisite questions are somehow off. The machine has not given him clearance to move forward. This man is the bottleneck. After his third attempt, two broad-shouldered TSA agents appear and stand beside the passenger.

"I think this thing is broken," the man informs them. The agents smile, shake their heads, take the man firmly by the elbow, and lead him away. The machine is not broken. The other passengers viewing this event don't know where the man is being led and express no concern. They understand only that an inconvenience has been moved from their path. They will catch their flight. Because the secondary search area is used only in rare circumstances and there is no backlog, even the man who has been pulled away will not be delayed too long—assuming, of course, the system determines he poses no legitimate threat.

An early version of the above-described program is already in place in strategically selected airports around the country (the metal detectors have not yet been moved out). The object of this screening is not luggage, exterior clothing, or even people's bodies, but rather people's innermost thoughts.

Today's computerized lie detectors take the form of Embodied Avatar kiosks. These watch eye dilation and other factors to discern whether passengers are being truthful or deceitful. No, the kiosk isn't going to do a cavity search, but it can summon an agent if it robotically determines you're just a bit too shifty to be allowed on a plane without an interview.[1]

Their functioning is based around the work of Dr. Paul Ekman, creator of one of the world's foremost experiments on lie detection, specifically how deception reveals itself through facial expression. Ekman's previous work has shown that with just a bit of training a person can learn to spot active deceit with 90 percent accuracy simply by observing certain visual and auditory cues—wide, fearful eyes and fidgeting, primarily—and do so in just thirty seconds. If you're a TSA agent and have to screen hundreds of passengers at a busy airport, thirty seconds is about as much time as you can take to decide if you want to pull a suspicious person out of line or let her board a plane.[2]

The biometric detection of lies could involve a number of methods, the most promising of which is thermal image analysis for anxiety. If you look at the heat coming off someone's face with a thermal camera, you can see large hot spots in the area around the eyes (the periorbital region). This indicates activity in the sympathetic-adrenergic nervous systems; this is a sign of fear, not necessarily lying. Someone standing in a checkpoint line with hot eyes is probably nervous about something.[3] The presence of a high degree of nervousness at an airport checkpoint could be considered enough justification for additional screening. The hope of people in the lie detection business is that very sensitive sensors placed a couple of inches away from a subject's face would provide reliable data on deception.

In 2006, the TSA began to experiment with live screeners who were being taught to examine people's facial expressions, mannerisms, and so on, for signs of lying as part of a program called SPOT (Screening Passengers by Observational Techniques).[4,5] When an airport-stationed police officer trained in "behavior detection" harassed King Downing, an ACLU coordinator and an African American, an embarrassing lawsuit followed. As Downing's lawyer John Reinstein told *New York Times* reporter Eric Lipton, "There is a significant prospect this security method is going to be applied in a discriminatory manner. It introduces into the screening system a number of highly subjective elements left to the discretion of the individual officer."[6]

Later the U.S. Government Accountability Office (GAO) would tell Congress that the TSA had "deployed its behavior detection program nationwide before first determining whether there was a scientifically valid basis for the program."

DARPA's Larry Willis defended the program before the U.S. Congress, noting that "a high-risk traveler is nine times more likely to be identified using Operational SPOT versus random screening."[7]

You may feel that computerized behavior surveillance at airports is creepy, but isn't a future where robots analyze our eye movements and face heat maps to detect lying preferable to one where policemen make inferences about intent on the basis of what they see? And aren't both of these methods, cop and robot, better than what we've got, a system that will deny someone a seat on a plane because her name bears a slight similarity to that of someone on a watch list? Probably the worst aspect of our airport security system as it currently exists is that evidence suggests we're not actually getting the security we think we are. As I originally wrote for the *Futurist*, recent research suggests that ever more strict security measures in place in U.S. airports are making air travel *less* safe and airports more vulnerable. So much money is spent screening passengers who pose little risk that it's hurting the TSA's ability to identify real threats, according to research from University of Illinois mathematics professor Sheldon H. Jacobson. Consider that for a second. We've finally reached a point where a stranger with a badge can order us to disrobe . . . in public . . . while we're walking . . . we accept this without the slightest complaint . . . *and it's not actually making us any safer.*

Our present system cannot endure forever. We won't be X-raying our shoes ten years from now. But what will replace it? What is the optimal way of making sure maniacs can't destroy planes while also keeping intercontinental air traffic on schedule?

The best solution, Jacobson's research suggests, is to separate the relatively few high-risk passengers from the vast pool of low-risk passengers long before anybody approaches the checkpoint line. The use of passenger data to separate the sheep from the goats would shorten airport screening lines, catch more threats, and improve overall system efficiency. To realize those three benefits we will all be asked to give up more privacy. We'll grumble at first, write indignant tweets and blog posts as though George Orwell had an opinion on the TSA, but in time we will exhaust ourselves and submit to predictive screening in order to save twenty minutes here or there. Our surrender, like so many aspects of our future, is already perfectly predictable. Here's why:

Our resistance to ever more capable security systems originates from a natural and appropriate suspicion of authority but also the fear of being found guilty of some trespass we did not in fact commit, of becoming a "false positive." This fear is what allows us to sympathize with defendants in a courtroom setting, and indeed, with folks who have been put on the wrong watch list and kept off an aircraft through no fault of their own. In fact, the entire functioning of our criminal justice system depends on all of us, as witnesses, jury members, and taxpayers, caring a lot about false positives. As the number of false positives decreases, our acceptance of additional security actually grows.

Convicting the wrong person for a crime is a high-cost false positive (often of higher cost than the crime). Those costs are borne mostly by the accused individual but also by society. Arresting an innocent bystander is also high cost, but less so. Relatively speaking, pulling the wrong person out of a checkpoint line for additional screening has a low cost but if you do it often enough, the costs add up. You increase wait time for everyone else (as measured by time that could be spent doing something else), and as Jacobson's model shows, it serves to erode overall system performance very quickly.

Now here's the tyranny of numbers: decrease the number of high-cost false positives and you can afford to make more premature arrests; bring *that* number down and you can afford more stop-and-frisks or security-check pat downs. Bring that number down again and the balance sheet looks like progress. The average citizen knows only that the system is improving. The crime rate appears to be going down; the security line at the airport seems to be moving faster. Life is getting better. If cameras, robots, big data systems, and predictive analytics played a part in that, then we respond by becoming more accepting of robots, cameras, and systems quantifying our threat potential when we're about to get on a plane. We grow more accustomed to surveillance in general, especially when submitting to extra surveillance has a voluntary component, one that makes submission convenient and resistance incredibly inconvenient. This is why, in the week following the disclosure of the massive NSA metadata surveillance program, a majority (56 percent) of Americans polled by Pew said they believed the tactics that the NSA was employing were acceptable. That's astounding considering that at the time, the media narrative was running clearly in the opposite direction.[8]

Another example of the opt-in surveillance state is the TSA's PreCheck program, which expedites screening for eligible passengers by rating their risk against that of the entire flying population. In order to be eligible for PreCheck, you're required to give the Department of Homeland Security a window into your personal life, including where you go, your occupation, your green card number if you're a legal alien, your fingerprints, and various other facts and tidbits of the sort that you could be forgiven for assuming the TSA had already (certainly the IRS has a lot of it). It's not exactly more invasive than a full body scan but in many respects it is more personal. Homeland Security uses the information it gets to calculate the probability that you might be a security threat. If you, like most people in the United States, are a natural-born citizen and don't have any outstanding warrants, you're not a big risk.

People who use TSA PreCheck compare it with being back in a simpler and more innocent time. But there's a downside, just as there is with customer loyalty programs at grocery stores. Programs such as PreCheck make a higher level of constant surveillance acceptable to more people. Suddenly, individuals who don't want to go along look extra suspicious. By definition, they are abnormal.

Security becomes faster, more efficient, and more effective through predictive analytics and automation so you should expect to be interacting with predictive screeners in more places beyond the X-ray line at the airport. But for computer programs, clearing people to get on a plane isn't as clear as putting widgets in a box. Trained algorithms are more sensitive than an old-school southern sheriff when it comes to what is "abnormal." Yet when a deputy or state trooper questions you on the side of the road, he knows only as much about you as he can perceive with his eyes, his ears, and his nose (or perhaps his dog's nose if his dog's at the border). Because the digital trail we leave behind is so extensive, the potential reach of these programs is far greater.

And they're sniffing you already. Today, many of these programs are already in use to scan for "insider threats." If you don't consider yourself an insider, think again.

Abnormal on the "Inside"

The location is Fort Hood, Texas. The date is November 5, 2009. It is shortly after 1 P.M.

Army psychiatrist Major Nidal Hasan approaches the densely packed Soldier Readiness Processing Center where hundreds of soldiers are awaiting medical screening. At 1:20 P.M., Hasan, who is a Muslim, bows his head and utters a brief Islamic prayer. Then he withdraws an FN Hertsal 5.7 semiautomatic pistol (a weapon he selected based on the high capacity of its magazine) and another pistol.[9]

As he begins firing, the unarmed soldiers take cover. Hasan discharges the weapon methodically, in controlled bursts. Soldiers hear rapid shots, then silence, then shots. Several wounded men attempt to flee from the building and Hasan chases them. This is how thirty-four-year-old police sergeant Kimberly D. Munley encounters him, walking quickly after a group of bleeding soldiers who have managed to make it out of the center. Hasan is firing on them as though shooting at a coven of quail that has jumped up from a bluff of tall grass. Munley draws her gun and pulls the trigger. Hasan turns, charges, fires, and hits Munley in the legs and wrists before she lands several rounds in his torso and he collapses. The entire assault has lasted seven minutes and has left thirteen dead, thirty-eight wounded.[10]

Following the Fort Hood incident, the Federal Bureau of Investigation, Texas Rangers, and U.S. chattering classes went about the usual business of disaster forensics, piecing together (or inventing) the hidden story of what made Hasan snap, finding the "unmistakable" warning signs in Hasan's behavior that point to the crime he was about to commit. After systematic abuse from other soldiers Hasan had become withdrawn. He wanted out of the military but felt trapped. Some of Hasan's superiors had pegged him as a potential "insider threat" years before the Hood shootings, but when they reported their concerns, nothing came of it. The biggest warning signal sounded in the summer of 2009 when Hasan went out shopping for very specific and nonstandard-issue firearms.[11]

The army had a lot of data on Hasan, much of which could have yielded clues to his intentions. The problem was that the army has a lot of data on everybody in the army. Some sixty-five thousand personnel were stationed at Fort Hood alone. The higher-ups soon realized that if they were to screen every e-mail or text message between soldiers and their correspondents for signs of future violence, it would work out to 14,950,000 people and 4,680,000,000 potential messages. Valuable warning signs of future insider threats were contained in those messages, which had been exchanged on systems and devices to which the army had access. But it was too much data for any human team to work through.

Not long after Fort Hood, army private Bradley Manning was arrested for giving confidential material to the Web site WikiLeaks, material that showed the United States was involved in killing civilians in Iraq. President Obama, who has proven to be exceedingly hard on whistle-blowing, responded with Executive Order 13587, which established an Insider Threat Task Force and mandated that the NSA and DOD each set up their own insider threat program.[12] DARPA issued a broad agency announcement on October 22, 2010, indicating that it was looking to develop a technology it called Anomaly Detection at Multiple Scales (ADAMS).[13]

The goal of this program is to "create, adapt and apply technology to the problem of anomaly characterization and detection in massive data sets . . . The focus is on malevolent insiders that started out as 'good guys.' The specific goal of ADAMS is to detect anomalous behaviors before or shortly after they turn," to train a computer system to detect the subtle signals of intent in e-mails and text messages of the sort that might have stopped the Fort Hood disaster, the Bradley Manning disclosure of classified information to WikiLeaks, or the Edward Snowden leak to the *Guardian* newspaper.

Varying bodies have differing definitions of what constitutes an "insider" in a military context but most agree that an insider is anyone with authorized access to any sensitive information that could be used against U.S. interests if disclosed improperly. What constitutes that sensitive information is a rather open-ended question but we know that it extends beyond the files, reports, or data has been officially labeled top secret.[14] For instance, the 2013 disclosures about the NSA PRISM system showed that several prominent Silicon Valley companies were forced to comply with NSA programs and orders from the secret Foreign Intelligence Surveillance Act (FISA) court. In that instance, an insider would include not just government workers or government contractors such as Edward Snowden but also any person at any of those private companies such as Google, Facebook, or Microsoft who simply knew of the existence of particular FISA orders.[15]

That broadness in the definition of both insider and outsider information is important for anyone concerned that an insider threat program could be abused. From the perspective of an algorithm, there is no meaningful difference between someone who is inside the military, inside the TSA PreCheck program, or inside Facebook. The same methods of anomaly detection can be applied to any observable domain.

We are all insiders.[16]

What are the telltale marks of a dangerous traitor? Some studies by military scholars list such seemingly benign traits as "lacks positive identity with unit or country," "strange habits," "behavior shifts," and "choice of questionable reading materials," adjectival phrases that describe virtually every American teenager since *Rebel Without a Cause*. The more provocative "stores ammunition" and "exhibits sudden interest in particular headquarters" seem of greater use but is something of a lagging indicator. And, naturally, it represents only one specific type of threat. In cyber-sabotage, attempting to gain access to a system or information source unrelated to your job function is often considered "abnormal" behavior. But how do you separate actionable abnormal from regular curiosity?[17]

No one at DARPA or any of the academic teams applying for the money is eager to discuss their research with reporters. Some of the most interesting work in predicting insider threats that has been made available to the public comes from a team of researchers led by Oliver Brdiczka and several of his colleagues at PARC. Instead of trying to pin down the traits associated with a treasonous personality, they sought to create a telemetric experiment where they could actually observe the threat develop. Here's another example of a simulation that would have been extremely costly a few years ago becoming cheap and relatively easy to perform thanks to more people living more of their lives online.[18]

Brdiczka and his colleagues looked at *World of Warcraft*, a massively multiplayer online game in which people develop a character, join teams called guilds, and go on quests that can last for days, during such time players effectively shun conventional hygiene practices or real-world contact with the opposite sex. Brdiczka had read the literature on how interpersonal dynamics, little exchanges between coworkers or between workers and supervisors, can predict workplace problems. *World of Warcraft* provided a perfect environment to telemetrically explore how group dynamics can turn a happy worker into a player who is willing to sabotage or steal from the members of his guild (a proxy for teammates). The researchers had each subject fill out a twenty-question survey to scan for the key personality traits of extroversion, agreeableness, conscientiousness, risk taking, and neuroticism, and they looked at the subjects' Facebook pages (and other social networking profiles) for similar personality clues. Then the researchers let them loose in the land of orcs and trolls.

Brdiczka and his team measured everything from which characters played more defensively to how quickly certain players achieved certain goals, what kinds of assignments they took, whether they gave their characters pets, how likely a subject was to shove a fellow player in front of a dragon to buy time to use a healing potion. Then they ran every verbal or text exchange between characters through a sentiment analysis algorithm to get a sense of how the subjects were communicating. In all, they looked for sixty-eight behavioral features related to how their players played the game. When they coupled those scores with the scores from the surveys and social network profiles, they found they could predict the players most likely to "quit" and thus sabotage their guild, within a six-month survey window, *with 89 percent accuracy.*

World of Warcraft functions as a useful proxy for all Internet interaction, and the government believes it has the right to access any of it. In 2012 the FBI created what it calls its Domestic Communications Assistance Center (DCAC) for the purpose of building back doors into the Internet and particularly into social networks, part of a sweeping Electronic Surveillance (ELSUR) Strategy. These online information collection devices join physical sensors, cameras, and scopes in the physical world and all of it contributes to an ever more revealing picture of our naked future.

We can pass legislation to keep the government from gathering information about us with certain methods, but as this surveillance infrastructure spreads and technology (particularly image recognition) improves, law enforcement won't need to use the most provocative, constitutionally questionable methods to get a credible picture of your present and future activities. When agents began putting GPS trackers on the undercarriage of cars to track the movement of suspects, they were sued and lost. Just a couple of years later the loss proved irrelevant. It turns out that tollbooth, streetlight, and security cameras all working together can track license plates across a city nearly as well as a GPS chip can broadcast a suspect's location.

While our risk of falling victim to violent crime in an American city is diminishing, the risk of being caught in the gears of an ever more powerful law and order apparatus is growing. This is not a trade-off many are willing to make. It's natural to dread a future in which our civilian power over law enforcement is diminished simply because we can't see what they can, because we have small amounts of information and they have complete recall of our digital trail. To be an uninformed populace is to be a disarmed one. When our local cops or our national security personnel are not only better armed but also exponentially more intelligent than are we, the chances for abuse of power increases and the challenge of reforming the system becomes greater. That's either a cause of concern or not, depending on your relationship with law enforcement.

These fears reflect reality, but not completely. In truth, some of us are much better informed than others. And the primary driver of the interconnected physical world is not government but garage entrepreneurs. The bigger threat to our privacy is not Big Brother; it's us.

The Return of Gordon Jones

Remember Guardian Watch from chapter 1? It was an Internet of Things service that allowed anyone with a video phone to stream live footage of a disaster to law enforcement, first responders, and the public. Not long after developing Guardian Watch, creator Gordon Jones realized that for the service to really flourish, to save the life of someone in an emergency, or particularly stanch a disaster affecting an entire city, it had to already be on phones, a lot of them. This presented the classic social start-up catch-22: the density problem. In order for Guardian Watch to become the next Foursquare of disaster response, it had to *already be* the Foursquare of disaster response; it had to have coverage, lots of users able to supply enough information and content to keep the app relevant.

Problem: the utility of Jones's creation during an emergency is obvious to anyone who has seen the demo, but nobody joins a social network while literally running for her life. Jones realized that network growth would depend largely on people adopting the service for reasons other than disaster preparedness. He rebranded the service (at the time called 911 Observer) as an enhanced neighborhood-watch network system: immediate help in emergencies—both real and imagined.

His first customer was the Richland County Sheriff's Department in Augusta, Georgia. In effect, Guardian Watch allows the department to crowd-source some of the more difficult aspects of evidence gathering.

This idea is not without precedent. One of former Texas governor Rick Perry's more creative legislative accomplishments was a program to digitally crowd-source border enforcement. The Texas Virtual Border Watch initiative enabled busybody constituents to monitor stretches of fence on the Texas-Mexico border from the comfort of their duct-taped La-Z-Boys, via live feed. The program was touted as a potential boon to taxpayers. The public was going to do for free what cost millions in pay to extra border guards.

The program failed for reasons having nothing to do with privacy and everything to do with why border patrolling is a hard job even on a good day. Watching a fence all day is *boring*. The site shut down in 2009 when, after an initial spike, traffic plummeted.

Guardian Watch allocates attentional resources to more interesting curated content, including but not limited to evidence of crime. Members can post pictures and videos from their phones into files such as "assault," "burglary," "domestic abuse," even "suspicious behavior." Much of the content uploaded thus far is of dubious value to law enforcement. One picture, marked "sexual," appears to show a nude couple enjoying coitus in a park . . . or a beached whale in a pasture . . . or a dinosaur. Many of the videos in the "suspicious behavior" file appear to show pant legs and shoes, all clearly shot by accident.

But these are the early days for the network, which Jones has marketed very selectively and which boasted thirty-nine hundred users as of summer 2012. If Guardian Watch can attract a following and funding, and can scale up to meet demand, if all the ducks and planets align themselves to favor him, his start-up or one like it could revolutionize not only emergency response but also law enforcement. To understand this potential, simply imagine a future in which geo-tagged pictures and video—images captured in the moment and digitally attached to a location, time, and person—take the place of unreliable witness testimony.

In addition to the clear privacy issues associated with this practice, there are questions of fecundity. The majority of content on social networking sites is personal and benign in nature, the daily annals of parenting and partying (sometimes both at once). A tweet or post about a suspicious person in your neighborhood is buried among a lot of other noise not relevant to law enforcement. The same problem hobbles most crime surveillance programs in urban areas. The United Kingdom has been experimenting with a camera program for years, one very similar to Texas Virtual Border Watch, but staffed by professionals who are paid to watch footage. Like the Texas program the vast majority of the footage is noise. The cost of sifting through it is high but you can automate it somewhat through algorithms. Guardian Watch represents a clear innovation in the way it enlists human beings to manually select evidence that's relevant to them.

Almost all the posted pictures pass through an intermediary before the content goes public. Jones wants to get rid of this step; a picture of a suspicious person in your neighborhood is really only valuable in real time. He also wants to further enhance the system with facial recognition capability, enabling it to tag people who show up in posted photos and videos automatically. That's fine if you trust your neighbors. But a vigilante could use real-time video of, say, a lone teen wandering nearby to quickly assemble a posse . . . or a lynch mob.

I pointed this out to Jones in the context of the Trayvon Martin case. Do we really want the George Zimmermans of the world to be more capable than they are now? He argued that the Martin case is a perfect example of the necessity of his system. Had a neighborhood resident been able to use Guardian Watch, Trayvon's father, Tracy, would have received a text or video about a suspicious person in his neighborhood in real time and seen Trayvon. He could send out a text alert to the group before George Zimmerman drew his gun.

I asked Jones about the prospect for more subtle forms of misuse. What's to stop someone from publicly tagging his neighbor as a domestic abuser or a terrorist? He answered that I have the power to post any number of unflattering things about my neighbors to any number of social networks and officially accuse anyone of domestic abuse with a five-minute phone call. But there's something of a social cost to posting that sort of content on Facebook. No similar social cost exists for posting the same material on Guardian Watch. It's the very purpose of the site.

The hope is that the citizen-policing system will self-police according to the same rules it uses to police others. Just as there exists a log of every Facebook tag and every domestic-disturbance call to the police, so every post on Guardian Watch creates data not just about the subject but also about the poster. Members who abuse the system lose influence.

At least that's the hope.

Naturally, anyone with a smartphone can already stream videos of people in her neighborhood to a Google+ circle or publicly through the Google+ Hangouts service, or to a specified group on Facebook. Guardian Watch has just taken the extra step of "friending" law enforcement and like-minded people on the user's behalf. Whether you see Jones's little start-up as a great way to improve public safety without increasing police budgets or as a lot of white people taking pictures of nonwhite people to make them nervous, Guardian Watch would exist without Gordon Jones.

Here's one of the more interesting examples: a Russian online newspaper called the *Village*, created an app provocatively named Parking Douche. It allows anyone with a smartphone to publicly shame a bad parker to that parker's neighbors, coworkers, and anyone in the neighborhood. To use it, you just take a picture of a car that's parked in a way that annoys you. The app then creates a banner ad with a picture of the car, license plate, and the name of the street. When people nearby (as determined by IP address) attempt to access online news though the *Village*'s Web site, they see the ad. It gets in their way. "Thanks to the IP address, only douches in your area will be highlighted, so all the offenders will be exposed to their colleagues' friends and neighbors," says the narrator in the demo video.[19]

What can we do to protect our privacy in a world where its value is falling faster than that of last year's cell phone? One creative if tongue-in-cheek proposal comes from British artist Mark Shepard whose Sentient City Survival Kit includes such items as a CCD-Me-Not umbrella studded with 256 infrared light-emitting diods (LEDs) to scramble the night vision of closed-circuit camera systems. My favorite item in the kit is the Under(a)ware, a set of undergarments that can detect RFID tags and vibrate to alert the wearer. "In the near future sentient shopping center, item level tagging and discrete data sniffing will become both pervasive corporate culture and a common criminal pastime," states a computerized voice on the demo video.

Unless our legal system becomes more transparent, accountable, and accessible we'll never feel certain that the people looking out for us won't abuse their power to persecute people who may technically be criminals but pose no real threat, such as pot smokers, prostitutes, and those who commit an act of trespass as part of a protest. How will we respond to this? Yes, we could put RFID tag readers in our underpants. *Alternatively*, we could decide to use surveillance and data to actually make the world safer and not abuse it. When you adopt the assumption that that's possible, opportunities open up.

If the bad news is the cops are going to have a better window into your career as a lawbreaker, the good news is that in the naked future you're more than just a suspect on her way to her next crime; you're a set of probabilities, potential costs, and potential benefits. The challenge for all of us now is to make the price of overzealous or discriminatory policing both high and conspicuous. The benefits of good policing must be more readily obvious as well. The social and public costs of pestering and prosecuting people for petty crimes should be visible to citizens, lawmakers, and police all at once. Before that happens we may have to settle for those costs becoming more transparent to law enforcement, where at least some departments or agencies will use them as part of their decision making. The same sort of technology that took away your privacy is beginning to provide just that opportunity.

The Microsoft Windows for Predictive Policing

The year is 2002. The Palo Alto–based online payment outfit PayPal has a big problem; Russian mobsters are defrauding the company to the tune of $3 to $4 million a month. PayPal's founder, Peter Thiel, and his coworker Joe Lonsdale realize they have to develop a system to better track the money moving through PayPal. Simply flagging individual transactions and users isn't enough. The infiltrators adapt far too quickly, setting up new locations and user identities before the old fake profiles grow cold. Thiel and Lonsdale know that if they can better chart how the money gets into the system and how it leaves, where it's spent and what it buys, then they can see whom it touches along the way.

They develop a new program to solve the problem. The Russian operation is broken. The fraud stops. It's at this point that they realize they've invented something they can spin out. They secure funding from the Central Intelligence Agency's investment arm, In-Q-Tel, and create Palantir Technologies, named for the seeing stones in *The Lord of the Rings*.

When I arrive at Palantir headquarters on a bright August day in 2012, I find halls filled with young programmers and graffiti-style murals. The only feature that distinguishes Palantir as a military and police service provider is the big-shouldered law enforcement types in the front lobby. I meet Courtney Bowman, a company spokesman and privacy expert. Bowman's got deep statistical-modeling experience as well as a background in philosophy. In his work at Palantir the second credential is just as useful as the first.

Palantir today is a platform to coordinate and organize files for different law enforcement purposes and agencies. It's like an operating system for classified or important information. The company itself doesn't do any evidence collection, snooping, or investigating. It simply offers software solutions to connect, centralize, and especially visualize information that may be distributed across a wide number of databases and players.

If you're a local law enforcement professional and you have a query about a suspect in a shooting, you can go to the Palantir interface on your desktop and find relevant records including arrests as well as suspect affiliations and even recent purchases. But the system doesn't give everyone the same access, as different clearance levels can exist across departments and agencies. If, however, you want to know where a particular record came from and who last updated it, the system can tell you that. If you want to see how two individuals may be connected to a single incident, crime, or transaction, the system can draw a map between the points of evidence.

"What Palantir can do," says Bowman, "is take those model outcomes or those hot-spot views, and, also, known information from criminal history records, from records management systems, from arrest records, and a multitude of other data sources that police legitimately have access to, and tie those all together into a picture of how the crimes, involving specific suspects or specific behavioral patterns, might play out."

For instance, say you're watching two suspects in a network. Person A is connected to person B through several affiliations. Person A makes a particular type of purchase, say, buying twelve rolls of toilet paper, before robbing a bank. The next day person B goes to a convenience store and buys twelve rolls of toilet paper. It's reasonable to infer he might be preparing to rob a bank. It's not enough to make an arrest but it does suggest an emerging pattern.

The practice of connection tracking, even when all that's being observed is correlation, is extremely fruitful in intelligence. In 2003, after months of trying to get information on Saddam Hussein's whereabouts from Hussein's senior officers and inner circle, the U.S. military used a social network mapping tool called i2 to chart the connections between his chauffeurs. This led them eventually to the farmhouse in Tikrit where Hussein was captured.[20]

Tracing the social network of a dictator during war is rather less controversial than analyzing the connections of millions of Americans. Yet this is what the U.S. government under the Obama administration has begun to do. The obscure National Counterterrorism Center (NCTC) routinely keeps personal transaction information, flight information, and other types of data on Americans who have neither been convicted nor are under suspicion of a crime. It does so for as long as five years under the vague auspice that it may be useful in some sort of investigation one day, even if that information isn't relevant to any operation at the time of collection.

The subjects of this transaction surveillance are people who have found their way into the Terrorist Identities Datamart Environment (TIDE), an enormous database of known terrorists, suspected terrorists, people who are loosely associated with suspected terrorists in some way (beekeepers, elementary school teachers, et cetera)—more than five hundred thousand links in all. The government has also given itself license to share the data across departments and even with other governments, despite the Privacy Act of 1974, which prohibits this sort of sharing.[21]

If legal, technical, and public relations costs of expanding surveillance remain as low as they are now, it's easy to imagine law enforcement considering a much broader array of connections and transactions worthy of monitoring.

But not all connection tracking in law enforcement is this creepy or controversial. For instance, say you're an investigator, and a roughneck from a particular gang, let's call him John Jet, is murdered. You know that the hit happened on a Friday night on disputed turf. Because this is gang related you're not just looking to solve the crime, you need to make an inference about the members of the rival gang (the Sharks) who have the highest probability of becoming victims of a retaliatory strike. Figuring this out may require sharing information across departments and even making inferences on the basis of correlations. But in this situation, you can deploy an antigang unit to a particular location at a particular time and do so without worrying about trampling on anyone's civil liberties.

While the NCTC may not need to consider itself accountable to the public, Palantir does. It's sort of like a beta tester. It plays a feedback role that is helping Palantir improve its system and make the system more valuable. As Bowman explains, "The government will make claims about collection of suspicious activity reports as being critical because you never know when that information is going to be useful. The privacy advocate will come back and say, 'Well, show me when this [piece of personal information on a subject] is actually useful. Give me hard metrics of why it's justifiable to hold on to this information.' If we can use the platform to demonstrate cases where this is useful, we can start to bridge the gap between these two communities and explain why this is valuable information."

The next step for Palantir, the product it wants to offer in the years ahead, is a model to allow law enforcement in the field to determine if the information it's got is good enough to bring a case. If you could structure data on court cases, as some folks at the Santa Fe Institute are currently trying to do, you could come up with a probability distribution for the likelihood of a court case's succeeding or failing. We are, in other words, fast approaching a future when it will be possible not only to see crimes in advance *but also see how the court case plays out.*

Today, we've convinced ourselves that we can't have improved public safety without giving up liberty. But perhaps in the future, children will see this trade-off as unnecessary, a failure of imagination. We've discounted the possibility that we can use public data and personal data in ways that empower individuals without making them feel uncomfortably exposed or more dangerous to one another. Get involved in how your local department uses or plans to use advanced analytics. Start a Facebook page that discusses how more involvement in how local police treat data is the trade-off we have to make for greater safety. You may get the brush-off, or you may be surprised to discover a bunch of smart public servants who are eager for more citizen participation. When police chiefs confront the reality of how income, employment, housing density, schooling, taxation, and even urban planning affect robbery, assault, and murder, they often start sounding less like cops and a lot more like sociologists.

The predictive policing program in Memphis, which has been in place longer than any other program of its type in the nation except New York's, has touched the lives of virtually everyone in the city. But it's attracted no complaints, no legal challenges, none of the controversy that has attached itself to other programs. Janikowski credits legwork for this. He and the deputy chiefs went to more than two hundred community meetings over the course of two years; they went through neighborhoods block by block to knock on doors, tell people what they were doing, and listen to concerns. Of these meetings, he says, "Whether there were five people there or five hundred, we did the same thing. We explained what we were doing, why we were doing it, what results we were hoping for. By going out there and telling folks, 'This is what's going on and why,' we never got the kinds of push back that I've heard from other cities."

A cynic would suggest that Janikowski surrendered a strategic advantage in doing so. He gave up too much information to the would-be perpetrators. Janikowski says being more generous with information and proactively reaching out to the public about the program, rather than just announcing its existence in a press release, is the reason the program continues to operate with the public's blessing. Public support is necessary if these programs are ever to reach their fullest potential.

The App That Saves You from Mugging

Crimes aren't just incidents that affect property values and insurance claims, they happen to people. Different people are in greater risk depending on various factors. These variables include circumstance: a drug kingpin versus a stay-at-home dad; situation, what you happen to be doing or are involved in at the time of the crime; and environment, where you are at what time of day.

These factors proceed upward along a continuum, says Esri's Mike King. The lower your risk, as determined by who and where you are, the less likely you are to experience crime at the hands of a stranger.

That means that if you *do* become a victim of crime, statistically speaking, someone you know is probably the culprit. Figuring out this sort of thing is an entire subfield in criminology called victimology, the study of how victims and perpetrators relate to one another. Although it's one of the most important aspects of criminal science (and the basis of the entire *Law & Order* franchise), victimology has never been formally quantified. We know that certain people are more likely to suffer crimes than others; a stay-at-home parent is much less likely to be the victim of a stabbing than a drug dealer. We also know that some areas are more dangerous than others. But we don't have a firm sense of exactly how these variables relate. Is a parent who is for some reason in a bad part of town more likely to be stabbed than a drug dealer in a church parking lot in the suburbs? And if so, how much more likely? The last real attempt to place values on some of these variables was in 1983.[22] It's time to try again.

Certainly, not every crime, perhaps not even most crimes, will fit some neat parameterization. Yet certain aspects of King's victimology continuum (such as victim's occupation and income) coupled with environmental factors (such as location, presence or absence of trees, and time of day) could be scored with enough data. They could make their way into a formula or algorithm that would output a particular victim-index score to anyone based on who she was, what she was doing, and where. Such an index would be useful for cops looking to establish a viable suspect pool after a crime occurs. But its real utility would be for individuals who could use their victimology score to edit their behavior.

Imagine that you are about to go out at night to deposit a check. You have a usual ATM that you go to in an area that you consider safe. You look at your phone and see a score reflecting the probability of getting mugged if you go to your regular spot or if you walk as opposed to drive. The score is up from just a couple of days ago; your neighborhood isn't as safe as you believed that it was.

Your future has been revealed to you, naked and shivering. You elect to go the ATM in the morning instead and to attend the next neighborhood meeting, maybe fix that broken window down the street. No crime is committed; no one has been harmed and no one's privacy has been violated. There is no victim.

CHAPTER 11
The World That Anticipates Your Every Move

THE year is 1992; the location, the Intel Corporation (INTC) campus of Santa Clara, California. Jeff Hawkins is giving a speech before several hundred of Intel's top managers. In the audience is Gordon Moore, the originator of Moore's law, and Andrew Grove, the CEO who transformed Intel into the most important computer parts manufacturer in the world. Hawkins arrives at Intel with a message of the future.

He tells his audience that people will soon be relying on something in their pocket as their primary computer, rather than the big desktop machines for which Intel manufactures chips. Furthermore, he says, these pocket computers will be so easy to use, so ubiquitous, and priced so affordably (he believed around $400 or so) that their popularity would dwarf that of conventional PCs.

Today, he describes the speech as the most poorly received talk he ever gave. The keynote was followed by a tense and brief Q&A session. "Usually when you give a talk you like people to say, 'Hey, that was great. I love that idea!' I didn't get any of that. I got like,

'Oh, well, that was interesting. I don't know if this really makes sense for us."

Hawkins became convinced of the bright future of mobile technology after several years tinkering with what would later be known as the world's first tablet PC, the great-great-grandfather of the iPad, a machine called the GRiDPad, which had a touch screen you used via a stylus. "People loved it," he recalls. "They were just immediately attracted to it. They went, 'Oh, I want to do things with this.' "

At $3,000 a unit, the GRiDPad was both an expensive and power-hungry piece of equipment not ready for the consumer marketplace. But Hawkins knew there was an audience for something like the device if he could make it smaller and much cheaper. A few months prior to his Intel talk, Hawkins created a company called Palm. Four years later, he would bring out the PalmPilot, the world's first digital assistant.

Time vindicated Hawkins completely. The great giants of the 1990s desktop-computing era, Dell, Gateway, and Hewlett-Packard, are looking as dour as old gray men in last night's rumpled evening suit, whereas mobile devices now comprise more than 61 percent of all the computers shipped.[1] None of the apps described in this book, either those real or those imagined, would be possible without Hawkins's insight that computers would become pocket-size and that computing would become something people did not just in labs, offices, or desks, but as they went about their lives. The mobile future that Hawkins saw decades ago is the origin of the naked future of today.

In 2012 I go to meet him at the headquarters of his company, Numenta (renamed Grok in 2013). The start-up, situated beside the Caltrain tracks in downtown Redwood, California, shares an office building with social networking company Banjo. It's a modest, even shabby office compared with the enormous campuses of Google, Facebook, and Apple that sit a few miles south on Route 82. A foosball table stands beyond the reception area where a scoreboard indicates that Jeff Hawkins, neuroscientist, inventor, reluctant futurist, is also dominating the rankings. "My team is playing a joke," says Hawkins. "I think I'm actually in last place." He plays only very rarely, when friends come to town. On the day I meet him, he has a match scheduled with Richard Dawkins in the afternoon.

The recently renamed company was founded in 2005 but didn't release its core product until 2012, a century in Silicon Valley years. Grok is a name taken from Robert A. Heinlein's 1961 novel *Stranger in a Strange Land*. It refers to a kind of telepathic commingling of thoughts, feelings, and fears: "Smith had been aware of the doctors but grokked that their intention was benign." The Grok is a cloud-based service that finds patterns in streaming (telemetric) data and uses those patterns to output continuous predictions.

Grok functioning is modeled on the neocortex. It's a hierarchical learning system made of columns and cells in the same way that the "new" part of the brain is made of neurons that connect to one another in layers, and then connect to neurons above or below.

What is the neocortex? It's the route of higher-order reasoning and decision making and the last part of the brain to evolve, hence the prefix "neo" (new). This new brain is present in all mammals but is particularly developed in humans. It doesn't look like much by itself, having no distinct shape. It's really more like a coat worn over the central brain. Because it's not well defined, the neocortex doesn't lend itself to easy summary. It's involved in too many different processes for that. But its composition is miraculous.

If you were to lay it out flat on a table and cut it open, you would discover six different types of neurons stacked on top of one another. They form a hierarchy much like the organizational chart of any company hierarchy. The bottom layer offices of Neocortex Inc. house the customer service reps; they collect customer feedback on a microscale. There are a lot of these types of neurons down there gathering data continuously from our hands, eyes, ears, and skin, typing it up, and sending it upstairs to the second floor where a different, less numerous set of neurons whittle down all the information they receive and pass the message upward again. This process repeats until the message hits the sixth floor, the executive offices. Up there, the top-level neurons are tasked with making decisions based on lots of incomplete data. It's a patchwork of reports, sensations, and impressions, a very incomplete picture of what's going on externally. The top-order neurons have to turn this info into something comprehensible and then send out an executable command in response. In order to get the picture to make sense, they have to complete it. To do that, our higher-order neurons draw from a storehouse of previously lived experience. They use that to extrapolate a pattern containing bits of recent external stimuli from the present, pieces already committed to working, and long-term memory from the past. That pattern is essentially a guess about what happens next, a prediction, a strange mix of fact and fiction.

"The brain uses vast amounts of memory to create a model of the world," Hawkins writes in his seminal 2004 book on the subject, *On Intelligence.* "Everything you know and have learned is stored in this model. The brain uses this memory-based model to make continuous predictions of future events. It is the ability to make predictions about the future that is the crux of intelligence."[2] In proposing this theory, Hawkins has also given rise to a new notion of prediction as a mental process that forms the very basis of what makes us human.

Each piece of data that enters the Grok system is processed by a different cell depending on its value and where it occurs in the data sequence. When data reaches a cell, let's call it cell A, the cell goes into an "active state" and establishes connections to other cells nearby—cell B if you will. So when cell A becomes active again, cell B enters a "predictive" state. If cell B becomes active, it establishes connections. If cell B becomes active again, the cells it establishes connections to enter a predictive state, and so on.

The operation of Grok is a bit less like a corporation, more like the old game of Battleship. You begin the game knowing only that your opponent's ships are somewhere on a board that looks like yours, so you call out coordinates. Scoring a direct hit gives you a hint about which nearby square to target next. If you hit the right sequence, you sink your opponent's battleship. Hawkins calls these little patterns "sparse distributed representations." They are, he says, how the brain goes about turning memories into future expectations. But in the brain these patterns are incredibly complex. The smell of Chanel No. 5 combined with a sound of distant chatter can trigger the memory of a particular woman in a black dress, moving through a crowded room. She's at a cocktail party. It's New Year's Eve. She is moving toward you and suddenly you look up, expecting to see her. Perhaps she is there again. Perhaps not. But someone is there now, someone wearing Chanel No. 5.

The Grok algorithm will not form unrequited longings toward women on New Year's Eve but its learning style is surprisingly human in comparison even to other neural networks. Grok experiences data sequentially, much the way humans do, as opposed to in batches, which is how most of our computer programs absorb information. The human brain simply can't use singular, static bits of data. We need a continuous stream. We don't recognize the notes of a song until we know exactly the order in which they follow one another as well as the tempo. We don't know we're touching alligator skin until our fingertips perceive the rough ridges rising and falling in quick succession. When looking at a picture, we may have an immediate recollection of the object depicted, but we won't know how it moves, what it does, its place in the world, until we see the object hurtling through time.

Grok, similarly, adjusts its expectations as quickly as it receives new information, without need of a human programmer to manually enter that data into the model. It's almost alive, in a somewhat primordial sense, constantly revising its understanding of the world based on a stream of sensory input. Says Hawkins, "That's the only way you're going to catch the changes as they occur in the world." He believes that hierarchical and sequential memory are the two most important elements in a truly humanistic artificial intelligence.

So far, Grok has performed a couple of proof-of-concept trials. The program is helping a big energy company understand what demand on the system will be every two hours, based on data that comes in minute by minute. They're working with a company that runs display ads. The client here is looking to figure out what different ad networks will soon be charging for cost per impression. This will allow the client to better plan when to run which ads. Grok is helping a windmill company in Germany predict when their machines will need repair. These are all rapid-fire predictions looking at the near future.

"That's exactly what brains do," says Hawkins. "[There are] several million, high velocity data streams coming into your brain. The brain has to, in real time, build a model of the data and then make predictions and act on the data. And that's what we need to do."

One consequence of this just-in-time predictive capability is that Grok doesn't retain all the data it receives. Data has a half-life. It's most valuable the moment it springs into being and depreciates exponentially; the faster the data stream, the shorter the half-life.

After millennia of rummaging about in the dark, creating signals that were lost to the ether the moment we brought them into existence, we've finally developed the capacity to form a permanent record of feelings, behaviors, and movements in a way that just a few years ago would have seemed impossible. We've developed the superpower of infinite memory. But in order to actually use it in a world where everything that was once noise is becoming signal, we must teach our machines how to forget.

"There's this perfect match between what brains do and what the world's data is going to be," says Hawkins. It's a point about which he's unequivocal. "Streaming is the future of data. It's not storing data someplace. And that's exactly what brains do."

What Hawkins doesn't mention, but very much knows, is that while it might have taken him decades to create a computer program capable of predicting the future, it took humanity a much longer time to develop the same capability.

Like any organic adaptive progression, human evolution was a clumsy and random process. Some 500 million years ago our pre-amphibian ancestors, inhabitants of the sea, were possessive of 100 million neurons. About 100 million years later, creatures emerged that processed sensory stimuli with but a few hundred million neurons. The tree of life diversified in branch and foliage. Competition bubbled up from the depths of the prehistoric swamp. Survival began to favor those organisms that could diversify with increasing rapidity. Great reptiles rose up and conquered the plains and jungles and these creatures, which we today condescendingly call stupid, had brains of several billion neurons. Their dominion was long and orderly compared with what ours has been since. But 80 million years after their departure, a brief intermission in terms of all that had come before, our mammalian ancestors grew, stood, developed the hominoid characteristics that we carry with us today, and evolved brains of 20 billion neurons; and this process continued to accelerate, building upon the pace it had established like an object falling through space, finally culminating 50,000 years ago in the human brain of 100 billion neurons and all the vanity, violence, wonder, delusion, and future gazing that it produces.

That saga stacks up pretty poorly to the evolution of mechanical intelligence, as documented by writer Steven Shaker in the *Futurist* magazine. The primordial computers of the 1940s were the size of houses but had only 200 or 300 bits of telephone relay storage. Fifteen years later, researchers at IBM were building machines with 100,000 bits of rotating magnetic memory, and then devices with hundreds of millions of bits of magnetic core memory just ten years after that. By 1975 many computers had core memories beyond 10 million bits. That figure increased again by a factor of ten in just ten years. By 1995 larger computer systems had reached several billion bits; and by the year 2000 it was not uncommon to see customized PCs with tens of billions of bits of random access memory.

Is computer data storage and processing in bits comparable with human information encoding in neurons? The former is electronic and travels literally at the speed of light; the latter, chemical based, is slower and more nuanced. Neurons combine and connect in ways that allow them to hold many memories at once. While we understand how to engineer memory in mechanical systems, we still have only the faintest notion of how the brain naturally does this so much better than computer systems. This is why the similarity between what Grok does when it makes a prediction and what the human brain does with the future is so significant.

You have a vision of the future in your head. In fact, you've harbored hundreds of thousands of different visions over the course of your lifetime. When your mind is at rest, it drifts into the future like a cottonwood sailboat pulled downstream by a gentle but persistent current. You dream of the future. Based on these visions, you take certain actions and avoid others. You govern your life in accordance with your hopes and fears of what the future will be even though every vision is an illusion.

What is the future? It is an organizational tool our species uses to delay gratification and stretch beyond our animal impulses. It is the idea of the future that allows us to build, save, invest, stifle impulses, and dedicate ourselves to objectives greater than the immediate. The future derives its functionality from its perceived changeability. We don't think about it without also imagining ourselves altering it.

Our belief in the future enabled us to sit and focus long enough to build stone tools required for hunting, then to plant crops and wait for harvest. It allowed for the creation of currencies, companies, colleges, retirement and health-care systems, and the like.

Here's a secret about the grandest of all human inventions: conjuring up the future is more than something we just do, it's the raison d'être for our most humanistic brain regions. The future is not a destination; it's a product of the brain, a product we evolved in order to make decisions in the present. Long before the capital-*F* Future became an imaginary destination, it was a tool that humankind evolved to turn memories into predictions about what will happen next in order to better our chances for survival in a very dangerous, wild world.

The work of Harvard neuroscientist Moshe Bar, University of

Ottawa researcher Jody Culham, Washington University professor Jeffrey Zacks, and others suggest that the cortical networks at play when we imagine the future—the ventromedial prefrontal cortex, the hippocampal formation, the medial parietal cortex, and particularly the medial temporal lobe—are the same ones associated with memory. We visualize the future in the same part of the brain that we use to recall the past and we exert similar amounts of effort—measured in hemoglobin flow—to do so.[3] Neurologically, the act of imagining a scenario is a lot like the act of remembering, says Bar. When these areas light up under fMRI, a prediction is born. That means our mental constructs of the future are a direct extension of our lived experience, a fact of neurological functioning that is central to the way we live and organize our lives.

Bar's research shows that we're also living in the future even when we don't realize it, when we're daydreaming. Our approach to it influences our every interaction, from our personality tics to the errors to which we are most susceptible. "One might wonder why our brain is investing energy in mind wandering, fantasizing and revisiting (and modifying) existing memories. I propose that a central role of what seems random thoughts and aimless mental simulations is to create 'memories.' Information encoded in our memory guides and sometimes dictates our future behavior," he writes.

As any attorney who has ever cross-examined a witness will tell you, memories aren't fixed. A person recalls things differently depending on the context she is in and what she's specifically trying to do. We alter our memories constantly depending on where we are, what we've just been through, and the circumstances under which we are being asked to remember. Bar's research provides a partial explanation for why this is. "Both real and simulated memories could be helpful later in the future by providing approximated *scripts* for thought and action" (emphasis added). How we remember predicts what we're going to do next.

We may actually *grieve* for the future as much as the past. Individuals who have devoted considerable time and mental resources to the visualization of a life spent with a particular person (in marriage) or to a particular life course (career) may show strong activity in the regions associated with planning. There may exist (again, theoretically) a measurable correlation between the amount of time specifically devoted to visualizing and planning a future with an individual and the amount of activity in these frontal regions after a breakup. In other words, you may be able to measure how traumatized someone is or will be by the dissolution of a relationship if you can measure the amount of time he's spent thinking about the future of that relationship. When we fall in love with a person, we're really falling for the future with them, an ungraspable dream.

A few years ago, across the United States from Bar's lab, Hawkins spearheaded a growing body of research to support the theory that the neocortex evolved expressly for the purpose of turning sensory data, in the form of lived experiences, into predictions. In many ways, the accomplishment he's most proud of is not a device or a company, it's the memory-prediction framework theory, a sort of unified theory for the brain purporting to explain why the brain, and specifically the neocortex, functions the way that it does.

Just as our gazing into the future takes a wide variety of forms, from strategizing to passive daydreaming, likewise the act of prediction serves a number of functions beyond just making decisions and plans. Prediction, in the brain, is also an act of learning.[4] We make predictions constantly as a means of testing the accuracy of what we've encoded and how we reexperience our memories. In fact, the predictions that have the most value to the brain as learning devices are the ones that are *wrong*.

Those experiences that bear out our expectations are much less likely to make their way into our permanent memory. We disregard them the same way we do boring episodes in our life, those little events that we were easily prepared for and that played out almost exactly how we anticipated that they would. The converse is true as well, says Bar. "Unpredictable incoming aspects that do not meet the possibilities offered by the top-down predictions can provide a signal both for attentional allocation as well as for subsequent memory encoding," he writes.[5]

When we're presented with an outcome that doesn't meet with our anticipations, our attention shifts. Our brains attempt to record the moment in as much detail as possible, not just the literal truth of what surprised us but other sensory data as well: the sound of the moment, the smell of it. We attempt to absorb the entire pattern. The feeling can be euphoric or it can match the clinical definition of trauma but the brain understands these experiences as learning, the acquisition of new information of possible use to survival.

We predict to learn but we also resist learning. In 2002 Duke neuroscientists Scott A. Huettel, Peter B. Mack, and Gregory McCarthy took sixteen volunteers and showed them a random series of shapes. When new incoming information violated the perceived pattern, the volunteers' brains showed significant activation, indicating surprise. Though the sequence was completely arbitrary, the fMRI showed that volunteers couldn't help but see—or rather invent—patterns in the sequence. No statistical fact ever feels more credible than our own experience.[6] Herein is where nature gets the last laugh; we're born predictors, but we're also bad predicators. We make up the future as we go along, get the answer wrong, and then convince ourselves we were right. This is why the inside view is so pernicious.

We evolved the brain, particularly the neocortex, to make guesses, but we did all that evolving over the course of millions of years. Mainstream science holds that the process of human evolution stopped some fifty thousand years ago, long before the dawn of human civilization. More important, this growing, changing, and adapting occurred in a natural setting where we had to hunt, scavenge, and avoid predators to survive.

That simple fact is essential to understanding how the future is becoming more visible. But our relationship to the future will always be psychological at its core. We engage the future through the act of prediction. What we now know about prediction is this:

1. Predictions in the brain come from lived experience.
2. Experience comes from processing sensory input—what we see, hear, and feel. If we can call the brain a type of computer (a common but incomplete analogy), this is data that our low-level neurons pick up, analyze, and then send to the higher-level neurons that make up the neocortex for processing and feedback.
3. The future changes depending on the amount of input and the manner in which it is processed.

So far, you've read about different devices that play a role in creating an anticipatory environment. Sensors embedded in the ground that detect P-waves enable us to predict S-waves; mobile devices we carry with us telegraph our future location based on where we've been. But the world that anticipates your every move will also be a wild and lifelike thing. It's a place full of moving parts, creatures that take actions based on what they can predict about your next move. They, too, are evolving the ability to anticipate the future, your future, on the basis of lived experience.

A few days after meeting Jeff Hawkins I journey to Esther's German Bakery in Los Galtos to meet with Paul Hofmann. Hofmann is the chief technology officer of Saffron Technology, a company that's also creating a continuous prediction platform based on neurological functioning. Hofmann is one of those European men who seem ageless. He has a thick mane of gray hair but his face is young and smooth. His accent is Bavarian but he sounds a lot like Hawkins in what he says.

"We will see the automation of cognitive thinking. It will happen like we have a pocket calculator," he predicts.

Saffron's product is a program that also builds models automatically based on incoming data; it also has a hierarchy, but the program emphasizes a mix of semantic and statistical reasoning. "We can analyze text, extract the semantics, and build context, counts, and then you can calculate entropy, relevance, matrices, and based on that, you can do pattern prediction," says Hofmann. At present, the market they're aiming for is one client, the government.

In addition to Hofmann, who was a vice president of research at SAP Labs, Saffron has a rather notable person on its board, John Poindexter, the former national security adviser and DARPA researcher. Following the 2001 9/11 terrorist attacks Poindexter pushed for a sweeping system of surveillance and terrorist targeting called Total Information Awareness (TIA). It involved analyzing e-mails, photos, and live camera footage to detect suspicious activity in real time. Critics called the plan a mass surveillance project. It was soon defunded and Poindexter was hounded out of his job.

Ironically, the portion of the proposal that cost Poindexter his position had little to do with surveillance. As part of the project, Poindexter wanted to set up a futures casino as a way to incentivize information discovery. Here's how it worked: if you were in the intelligence community (an umbrella term that applies not just to working spies but to former spies, informants, and analysts, an internationally dispersed web of thousands of people) and you had a suspicion about a crime or a terrorist plot, you could wager on the likelihood of the event's outcome and make money. According to Poindexter's theory, analysts or informants would be more likely to be forthcoming with information of value if they could profit by it directly and anonymously.

Poindexter's prediction market represented a distinct innovation over the way various federal agencies had gone about the task of collecting and coordinating information before the 9/11 terrorist attacks. Poindexter showed that his method was more accurate and effective than conventional information gathering. It also proved to be a political flash point.

When word of the prediction market project hit the public, Senator Bryan Dorgan (D-ND) claimed that because the program rewarded people for having inside information without asking where they got it, it might well *cause* the very attacks it was invented to predict. A full investigation of Poindexter's Information Awareness Office (IAO) within DARPA followed, which led to the public revelation of domestic spying activity on the part of the office. Funding was cut and John Poindexter submitted his resignation to DARPA director Anthony Tether on August 13, 2003.

If you look at the TIA proposal today, it appears much less controversial than it did a decade ago. Several of the information-gathering methods proposed are indeed in use by law enforcement and counterterrorism agencies around the world. Dozens of private online prediction markets have sprung up, allowing people to bet on any number of world events.

TIA is effectively live.

I ask Hofmann about this, whether he shares the privacy concerns of people who argue that we were moving toward a massive surveillance state.

"Privacy is a blip on the radar of history," he tells me.

He takes a bite from his thick, buttery pastry, sips from his iced tea, and calmly continues.

"Until I was eight years old, I lived in a small village in the Austrian countryside. Then my parents moved me to Vienna. I experienced, for the first time, anonymity. Nobody knew me. Now, the world is a global village. Everyone knows everyone again."

In 1998, roughly ten years after Mark D. Weiser coined the term "ubiquitous computing" to describe a world in which the objects in our environment became passive listeners, futurist and science-fiction author David Brin published a nonfiction book titled *The Transparent Society*, which discusses the future effects of ubiquitous computing. The first chapter poses two alternative futures. One saw the Orwellian vision of total state surveillance. Empowered and active citizens dominated the second scenario and they used the new capabilities of the awakened environment to check and humble those who would call themselves guardians of the public order.

"In the Information Age to come, cameras and databases will sprout like crocuses—or weeds—whether we like it or not. Over the long run, we as a people must decide. Can we stand living exposed to scrutiny . . . our secrets laid open . . . if in return we get flashlights of our own, that we can shine on anyone who might do us harm? Even the arrogant and strong? Or is an illusion of privacy worth any price, even surrendering our own right to pierce the schemes of the powerful?"[7]

Today, we have the ability to build these flashlights of our own by taking charge of our own data through the services and strategies outlined in this book. We may not want to literally install sensors in the sewer system like Leif Percifield or document the moving winds of Fukushima, but the same tools that others use to detect change are rapidly becoming more available to all. Whether we take advantage of these opportunities or not, there will be light.

The foretold information age has arrived, the illusion of privacy has been revealed as such. We voluntarily lay our lives, and those of our loved ones, before the eyes of the world and feel slighted when the masses treat our deepest secrets as irrelevant. For many of us, self-revelation has become a strange form of compulsion. For the rest of us, it's become an accident that happens ever more frequently.

We would do well to remember that the most interesting, hopeful, relevant, and lifesaving work in creating a smarter world is being done not by multinationals, not by Big Brother, but by regular people: garage entrepreneurs, activists, and hackers. There is no one, not in any government office or corporate suite, who fully understands the current global transformation toward a naked future.

Gus Hunt, the chief technical officer for the CIA, a man closer to the future of technology than nearly anyone, is among the most ambivalent about our ever smarter environment. At a meeting of data scientists in downtown Washington in early 2012, he spent a long time discussing recent innovations that caught him off guard including a refrigerator that could predict when its owner was about to run out of milk and wheelchairs controlled by electromagnetic bursts from the user's motor cortex. He saw these breakthroughs as inexorably linked to big data and the future. "These are an additional accelerant to the thing we called big data. The sensors themselves will be talking about everything going on on the planet." He spoke excitedly, a man privileged to have access to the cutting edge of technology. But he is as small and insignificant in the face of these changes as anyone else. "My dystopian nightmare is a world where this data is floating around and you go to buy your morning coffee, you swipe your debit card, and the health company texts you and says you have high blood pressure. And then the government charges you a tax."

If Gus Hunt can't control this future, if he is not the government, who is?

We are on the verge of a new world where our systems perceive the present and the future in much the same way we learned to extrapolate the future millions of years ago. But whereas we were once limited to the sensory perceptions of a singular organism, the world we are now creating is one where experience, reduced to signals, can be shared universally, unconsciously, immediately, permanently, or temporarily, depending on what sort of future we're seeking to access. We will approach these myriad futures with a purpose and certainty that our ancestors would never have thought possible. In this way we will come to resemble gods. Yet we will feel increasingly powerless against the tide of transparency rendering this planet in a new form as surely as the movement of glaciers carved our canyons and valleys.

The Obituary for Privacy

Today, the data we create in our comings and goings is mostly separate from the information we post on Facebook, Twitter, or Google. This division is temporary. The incentives for removing the barriers are already larger than the reasons for keeping them in place.

For instance, after centuries of trying to use math to figure out who should be with whom, only very recently have we developed the means to measure on a day-by-day, second-by-second basis the billion bits, small information exchanges, conversations, moments of happiness, awe, and disappointment that make up a relationship. We can use this to make better decisions, better avoid affairs of the heart that are doomed from the start and strengthen those that we want to keep.

What are some of the dangers of our new predictive age? Activist and author Eli Pariser wrote of some of them in his book *The Filter Bubble*. The title refers to a type of "informational determinism," the inevitable result of too much Web personalization. The filter bubble is a state in which "what you've clicked on in the past determines what you see next—a Web history you're doomed to repeat. You can get stuck in a static, ever-narrowing version of yourself—an endless you-loop."

Google and Facebook are perhaps the two most obvious examples of companies using your data to better predict your behavior. But you can always opt out of using Facebook, as millions already have. And while cutting Google out of your life isn't as easy as it was a decade ago, there are ways to use Google anonymously. An arguably more pernicious threat is posed by systems and companies that are using our information to make predictions about us without our even knowing.

As we become participants in systems, networks, and communities where data collection plays a role; as we encounter more apps, programs, and platforms that need our data to run; predictability improves as privacy vanishes, a consequence of computers making record keeping and record sharing easier and cheaper.

If I can impart one piece of advice in reading this book, it's this: we will not win by shaking our fist in the air at technology. A better solution is to familiarize ourselves with how these tools work; understand how they can be used legitimately in the service of public and consumer empowerment, better living, learning, and loving; and also come to understand how these tools can be abused.

I'm not yet entirely comfortable in this world. I, too, grow cold at the thought of robots peering down at me anticipating my location in the next few seconds, few minutes, perhaps years from now, or cops in patrol cars looking at me with a narrowed eye, seeing a 10 percent probability of check bouncing or an 80 percent chance of committing a parking violation in the next hour. All I do know is that my discomfort won't stop the winds that are revealing to the world my intentions, purchases, illnesses, hopes, and fears, my life more clearly for what it is. These capabilities that are emerging from Silicon Valley and Washington, from labs, offices, and garages, are what the military refers to as "force multipliers," like mustard gas during World War I or night vision goggles during Desert Storm. The thing about force multipliers is that once they're out of the box, they don't go back in.

In a world where everyone has the superpower to see the future, but some can see further, "super" will become meaningless. The naked future will bring with it not just safer cities, smarter students, better movies, healthier bodies, and wider awareness but also new frustrations, inconveniences, and forms of unfairness, to which we will respond with insult. This is what humans do with technological advancement. When given the opportunity to travel five hundred miles per hour thousands of feet in the air, over mountains, oceans, and skyscrapers, we grumble that the people in the front of the mechanical flying marvel have more legroom. Knowledge, widely distributed, does not automatically produce happiness even if it does cure ignorance.

Although the future will soon feel very different, it will remain fundamentally what it has always been: a state that we create through our decisions but cannot in fact experience. It is a picture we can see with ever greater clarity, but it is an image we will change, an image we are changing. By the time you finish reading this book, it will have changed again.

ACKNOWLEDGMENTS

I sometimes tell people that I have the best job in the world because I am privileged to meet and interact with the planet's smartest folks. That was particularly the case here. I am especially grateful to my interview subjects, including Stephen Wolfram, Adam Sadelik, John Wilbanks, Sacha Chua, Michael Paul, Jehoshua Eliashberg, Peter Norvig, Gordon Jones, and Jeff Hawkins among the various other scientists, entrepreneurs, and pioneers who took time away from important projects to answer my questions. My sincere hope is that I did well by your work, even in those instances where I felt the need to express some moral reservation.

I must also thank my editor, Niki Papadopoulos, for her focused and clear insight, her tremendous patience, her consistency, and her delicate honesty. This book likewise would not have been possible without the expert advocacy and superb council of my agent, Loretta Barrett, and her colleague Nick Mullendore at Loretta Barrett

Books. I am truly thankful to my publisher, Adrian Zackheim, for his incredible support of this work, and everyone at Current Books.

I would like to thank Cynthia Wagner for making me a better writer, a better editor, for being an exceptional boss, a true friend, and for allowing me to pursue those stories, angles, or wild geese that I felt compelled to chase. I would like to thank Edward Cornish for his mentorship, inspiration, and vision, and Jeff Cornish, Tim Mack, the World Future Society, and the *Futurist* magazine for affording me every opportunity to explore the future and all that it can be. My father, sister, stepmother, mother-in-law, and father-in-law can be relied on to make a big deal out of even my smallest successes and to forgive the fact that I seem never to be available as I am always typing away at something.

Most important, I must acknowledge my wife, Beth, for her unwavering partnership, generosity, and love.

Life is a long lesson in humility, said Sir James Matthew Barrie. He could also have been talking about what this book has been for me. I have many reasons to feel humble.

NOTES

INTRODUCTION

1. *The Oxford English Dictionary*, 2nd ed. (New York: Oxford University Press, 1989), 20–21.
2. "Big Data, for Better or Worse: 90% of World's Data Generated over Last Two Years," *ScienceDaily*, May 22, 2013, http://www.sciencedaily.com/releases/2013/05/130522085217.htm.
3. Patrick Tucker, "Has Big Data Made Anonymity Impossible?" *MIT Technology Review*, May 7, 2013, http://www.technologyreview.com/news/514351/has-big-data-made-anonymity-impossible.
4. Ibid.

CHAPTER 1: NAMAZU THE EARTH SHAKER

1. Mitsuyuki Hoshiba et al., "Outline of the 2011 Off the Pacific Coast of Tohoku Earthquake (M w 9.0)—Earthquake Early Warning and Observed Seismic Intensity," *Earth Planets Space* 63 (Sept. 27, 2011): 547–51.
2. *Earthquake and Tsunami Alarm Systems of Japan, March 11, 2011*, 2011, http://www.youtube.com/watch?v=24KfBwkMw_M&feature=youtube_gdata_player.
3. Rachel A. Grant, "Predicting the Unpredictable: Evidence of Pre-Seismic Anticipatory Behaviour in the Common Toad," *Journal of Zoology*, 281, no. 4 (Aug. 2010): 263–71, doi:10.111.
4. Rachel A. Grant et al., "Ground Water Chemistry Changes Before Major Earthquakes and Possible Effects on Animals," *International Journal of*

Environmental Research and Public Health 8, no. 12 (June 1, 2011): 1936–56, doi:10.3390/ijerph8061936.

5. Nate Silver, *The Signal and the Noise: Why So Many Predictions Fail—But Some Don't* (New York: Penguin Press, 2012), 148, https://catalyst.library.jhu.edu/catalog/bib_4315599.
6. Mark D. Weiser and John Seely Brown "The Coming Age of Calm Technology," Xerox PARC, October 5, 1996, accessed Oct. 10, 2013. http://www.ubiq.com/hypertext/weiser/acmfuture2endnote.htm.
7. Joe Burton, "Wearable Devices to Usher in Context-Aware Computing," *ZDNet*, accessed Jan. 20, 2013, http://www.zdnet.com/blog/emergingtech/wearable-devices-to-usher-in-context-aware-computing/3276.
8. Ben Gruber, "First Wi-Fi Pacemaker in US Gives Patient Freedom," *Reuters*, August 10, 2009, http://www.reuters.com/article/2009/08/10/us-pacemaker-idUSTRE5790AK20090810.
9. Courtney Boyd Myers, "Developing a Wireless System for Detecting Explosive Devices," *Next Web*, accessed Jan. 20, 2013, http://thenextweb.com/shareables/2011/11/23/developing-a-wireless-system-for-detecting-explosive-devices.
10. Rich Smith, "Shocking Way Electric Utilities Are Making Us Pay for the Smart Grid," *DailyFinance*, accessed Jan. 20, 2013, http://dailyfinance.com/2012/05/15/electric-utilities-smart-grid-greed.
11. Russell Brandom, "Can You Find Me Now? How Carriers Sell Your Location and Get Away with It," *Verge*, Apr. 9, 2013, http://www.theverge.com/2013/4/9/4187654/how-carriers-sell-your-location-and-get-away-with-it.
12. Yves-Alexandre de Montjoye et al., "Unique in the Crowd: The Privacy Bounds of Human Mobility," *Scientific Reports* 3 (Mar. 25, 2013), doi:10.1038/srep01376.
13. Kathryn Zickuhr, "Three-quarters of Smartphone Owners Use Location-Based Services," Pew Research Center's Internet & American Life Project, May 11, 2012, http://pewinternet.org/Reports/2012/Location-based-services.aspx.
14. Boonsri Dickinson, "Dave Morin Explains Why Britney Spears Visited Path," *Business Insider*, Mar. 23, 2012, http://www.businessinsider.com/this-is-why-britney-spears-went-to-path-2012-3.
15. "Startup Path Bids to Be 'Anti-Social Network,'" *Economic Times*, accessed Jan. 20, 2013, http://articles.economictimes.indiatimes.com/2010-11-16/news/27590386_1_social-network-facebook-limits.
16. Arun Thampi, "Path Uploads Your Entire iPhone Address Book to Its Servers," mclov.in, February 8, 2012, http://mclov.in/2012/02/08/path-uploads-your-entire-address-book-to-their-servers.html.
17. Judea Pearl, *Causality: Models, Reasoning, and Inference*, 2nd ed. (New York: Cambridge University Press, 2009), https://catalyst.library.jhu.edu/catalog/bib_3508574.
18. Adam Sadilek, Henry Kautz, and Jeffrey P. Bigham, "Finding Your Friends and Following Them to Where You Are," in *Proceedings of the Fifth ACM International Conference on Web Search and Data Mining*, WSDM 2012 (New York: ACM, 2012), 723–32, doi:10.1145/2124295.2124380.

19. Adam Sadilek and John Krumm, "Far Out: Predicting Long-term Human Mobility," *Twenty-Sixth AAAI Conference on Artificial Intelligence*, 2012, http://www.aaai.org/ocs/index.php/AAAI/AAAI12/paper/view/4845.

CHAPTER 2: THE SIGNAL FROM WITHIN

1. Kashmir Hill, "Fitbit Moves Quickly After Users' Sex Stats Exposed," *Forbes*, July 7, 2011, http://www.forbes.com/sites/kashmirhill/2011/07/05/fitbit-moves-quickly-after-users-sex-stats-exposed.
2. Susannah Fox and Maeve Duggan, *Tracking for Health* (Pew Internet & American Life Project, Jan. 28, 2013), http://www.pewinternet.org/Reports/2013/Tracking-for-Health/Summary-of-Findings.aspx.
3. Kevin Kelly, "Self-Tracking Devices Are Sort of Main Stream Now. Just . . . ," Jan. 15, 2013, Google+, https://plus.google.com/+KevinKelly/posts/139fExY8d1N.
4. Graydon Carter, "Recording Artists and Sadistic Chefs," *Vanity Fair*, Feb. 1, 2013, http://www.vanityfair.com/magazine/2013/02/graydon-carter-quantified-self.
5. *The Autobiography of Benjamin Franklin*, chapter 8, EarlyAmerica.com, accessed Jan. 20, 2013, http://www.earlyamerica.com/lives/franklin/chapt8.
6. Justin Kruger and David Dunning, "Unskilled and Unaware of It: How Difficulties in Recognizing One's Own Incompetence Lead to Inflated Self-Assessment," *Journal of Personality and Social Psychology* 77, no. 6 (1999): 1121–34.
7. Michael Mauboussin, "Smart People, Dumb Decisions," *Futurist*, Apr. 2010.
8. Terry Grossman, "Ray Kurzweil's Plan for Cheating Death," *Futurist*, Apr. 2006.
9. Stephen Wolfram, "The Personal Analytics of My Life," *Stephen Wolfram Blog*, Mar. 8, 2012, http://blog.stephenwolfram.com/2012/03/the-personal-analytics-of-my-life.
10. Jennifer R. Piazza et al., "Affective Reactivity to Daily Stressors and Long-term Risk of Reporting a Chronic Physical Health Condition," *Annals of Behavioral Medicine: A Publication of the Society of Behavioral Medicine* (Oct. 19, 2012), doi:10.1007/s12160-012-9423-0.
11. Sara LaJeunesse, "Reactions to Everyday Stressors Predict Future Health," Nov. 2, 2012, http://live.psu.edu/story/62452.
12. Lisa M. Vizer and Andrew Sears, "Detecting Cognitive Impairment Using Keystroke and Linguistic Features of Typed Text: Toward an Adaptive Method for Continuous Monitoring of Cognitive Status," in *Proceedings of the 7th Conference on Workgroup Human-Computer Interaction and Usability Engineering of the Austrian Computer Society: Information Quality in e-Health*, USAB 2011 (Berlin, Heidelberg: Springer-Verlag, 2011), 483–500, doi:10.1007/978-3-642-25364-5_34.
13. Kate Kaye, "There's Data in That Toothbrush (And Lots of Other Products, Too)," *Ad Age*, May 20, 2013, http://adage.com/article/dataworks/toothbrushes-pill-packages-record-consumer-data/241557.
14. Tyler McCormick, Cynthia Rudin, and David Madigan, *A Hierarchical Model for Association Rule Mining of Sequential Events: An Approach to Automated Medical Symptom Prediction*, SSRN Scholarly Paper (Rochester,

NY: Social Science Research Network, Jan. 8, 2011), http://papers.ssrn.com/abstract=1736062.
15. Mats G. Hansson et al., "Ethics Bureaucracy: A Significant Hurdle for Collaborative Follow-up of Drug Effectiveness in Rare Childhood Diseases," *Archives of Disease in Childhood* 97, no. 6 (June 1, 2012): 561–63, doi:10.1136/archdischild-2011-301175.

CHAPTER 3: #SICK

1. "Phenotype," *Wikipedia*, Dec. 21, 2012, http://en.wikipedia.org/w/index.php?title=Phenotype&oldid=528449757.
2. Rita Ruben, "Flu Shot Not as Effective as Thought (But Get One Anyway)," *NBC News*, Oct. 25, 2011, http://vitals.nbcnews.com/_news/2011/10/25/8484876-flu-shot-not-as-effective-as-thought-but-get-one-anyway.
3. "Seasonal Influenza (Flu)—Key Facts About Seasonal Flu Vaccine," CDC, accessed Jan. 21, 2013, http://www.cdc.gov/flu/protect/keyfacts.htm.
4. "EpiFlu—Global Initiative on Sharing All Influenza Data," accessed Jan. 21, 2013, GISAID, http://platform.gisaid.org/epi3/frontend#11f98c.
5. National Center for Biotechnology Information, accessed Jan. 21, 2013, http://www.ncbi.nlm.nih.gov.
6. "First H3N2 Variant Virus Infection Reported for 2012," CDC, Apr. 12, 2012, http://www.cdc.gov/flu/spotlights/h3n2v-variant-utah.htm.
7. Declan Butler, "Flu Surveillance Lacking," *Nature* 483, no. 7391 (Mar. 28, 2012): 520–22, doi:10.1038/483520a.
8. Yi-Mo Deng, Natalie Caldwell, and Ian G. Barr, "Rapid Detection and Subtyping of Human Influenza A Viruses and Reassortants by Pyrosequencing," *PLoS ONE* 6, no. 8 (Aug. 19, 2011): e23400, doi:10.1371/journal.pone.0023400.
9. E. K. Subbarao, W. London, and B. R. Murphy, "A Single Amino Acid in the P2B Gene of Influenza A Virus Is a Determinant of Host Range," *Journal of Virology* 67, no. 4 (Apr. 1, 1993): 1761–64.
10. Sander Herfst et al., "Airborne Transmission of Influenza A/H5N1 Virus Between Ferrets," *Science* 336, no. 6088 (June 22, 2012): 1534–41, doi:10.1126/science.1213362.
11. Brendan Maher, "Bird-Flu Research: The Biosecurity Oversight," *Nature* 485, no. 7399 (May 24, 2012): 431–34, doi:10.1038/485431a.
12. Tyler Kokjohn and Kimbal Cooper, "In the Shadow of Pandemic," *Futurist*, Oct. 2006.
13. Patrick Tucker, "Catching a Pandemic, Online," *Futurist*, June 2013, http://www.wfs.org/futurist/2013-issues-futurist/may-june-2013-vol-47-no-3/catching-pandemic-online.
14. Eric Topol, *The Creative Destruction of Medicine* (New York: Basic Books, 2011), 265.
15. Marcel Salathé et al., "A High-resolution Human Contact Network for Infectious Disease Transmission," *Proceedings of the National Academy of Sciences* 107, no. 51 (Dec. 21, 2010): 22020–25, doi:10.1073/pnas.1009094108.
16. Nicholas Christakis and James Fowler, *Connected: How Your Friends' Friends' Friends Affect Everything You Feel, Think, and Do* (New York: Little Brown, 2011), 19.

17. Michael R. Moser et al., "An Outbreak of Influenza Aboard a Commercial Airliner," *American Journal of Epidemiology* 110, no. 1 (July 1, 1979): 1–6.
18. M. J. Paul and M. Dredze, "A Model for Mining Public Health Topics from Twitter," *HEALTH* 11 (2011): 6–16.
19. Adam Sadilek, Henry Kautz, and Vincent Silenzio, "Modeling Spread of Disease from Social Interactions," in *Sixth International AAAI Conference on Weblogs and Social Media*, 2012, http://www.aaai.org/ocs/index.php/ICWSM/ICWSM12/paper/view/4493.
20. "Feeling Sick Makes Us Less Social Online Too," Brigham Young University Media Page, Mar. 25, 2013, http://news.byu.edu/archive13-mar-socialhealth.aspx.

CHAPTER 4: FIXING THE WEATHER

1. Wolfgang Wagner, "Taking Responsibility on Publishing the Controversial Paper 'On the Misdiagnosis of Surface Temperature Feedbacks from Variations in Earth's Radiant Energy Balance' by Spencer and Braswell, Remote Sens. 2011, 3(8), 1603–1613," *Remote Sensing* 3, no. 12 (Sept. 2, 2011): 2002–4, doi:10.3390/rs3092002.
2. Bill Chameides, "Global Warming: Fox News Separates Fact from Fiction," *Green Grok*, accessed Jan. 20, 2013, http://blogs.nicholas.duke.edu/thegreengrok/bastardi.
3. Roy Spencer, "Editor-in-Chief of Remote Sensing Resigns from Fallout over Our Paper," Royspencer.com, Sept. 2, 2011, http://www.drroyspencer.com/2011/09/editor-in-chief-of-remote-sensing-resigns-from-fallout-over-our-paper.
4. Peter Lynch, *The Development of Atmospheric General Circulation Models: Complexity, Synthesis and Computation* (Cambridge: Cambridge University Press, 2011), 3–17.
5. James R. Fleming, "Sverre Petterssen, the Bergen School, and the Forecasts for D-day," in *Proceedings of the International Commission on History of Meteorology*, 2004, http://www.scribd.com/doc/78949967/Sverre-Petterssen-and-the-Forecasts-for-D-Day.
6. Lewis L. Strauss, *Men and Decisions*, 1st ed. (New York: Doubleday, 1962).
7. Mary Bellis, "The History of the ENIAC Computer," About.com, accessed Jan. 20, 2013, http://inventors.about.com/od/estartinventions/a/Eniac.htm.
8. James Rodger, *Fixing the Sky: The Checkered History of Weather and Climate Control* (New York: Columbia University Press, 2010).
9. George Dyson, *Turing's Cathedral: The Origins of the Digital Universe*, 1st ed. (New York: Pantheon Books, 2012), 171.
10. "Vision Prize Expert Poll Findings," *Vision Prize*, Mar. 31, 2012, http://visionprize.com/results.
11. Kieran Mulvaney, "Climate Change, Bikes and Black Helicopters," *DNews*, Feb. 16, 2012, http://news.discovery.com/human/climate-change-bike-sharing-and-black-helicopters-120216.htm.
12. Lyle Scruggs and Salil Benegal, "Declining Public Concern About Climate Change: Can We Blame the Great Recession?" *Global Environmental Change* 22, no. 2 (May 2012): 505–15, doi:10.1016/j.gloenvcha.2012.01.002.

13. Christine Parthemore and Will Rogers, *Blinded: The Decline of US Earth Monitoring Capabilities and Its Consequences for National Security* (Center for a New American Security), accessed Jan. 20, 2013, http://www.cnas.org/files/documents/publications/CNAS_Blinded_ParthemoreRogers_0.pdf.
14. "Obama Warns GOP Budget Would Make Weather Prediction Less Accurate," *Real Clear Politics*, accessed Jan. 20, 2013, http://www.realclearpolitics.com/video/2012/04/03/obama_warns_gop_budget_would_make_weather_prediction_less_accurate.html.
15. "Stanford's Entrepreneurship Corner: David Friedberg, The Climate Corporation—Entrepreneurship Gives Life Meaning," accessed Jan. 20, 2013, *ECcorner*, http://ecorner.stanford.edu/authorMaterialInfo.html?mid=2789.
16. John Roach, "Insuring Against Extreme Weather," *NBC News*, accessed Jan. 20, 2013, http://www.nbcnews.com/technology/futureoftech/insuring-against-extreme-weather-120060.
17. Ashlee Vance, "Climate Corp. Updates Crop Insurance via High Tech," *Businessweek*, Mar. 22, 2012, http://www.businessweek.com/articles/2012-03-22/climate-corp-dot-updates-crop-insurance-via-high-tech.
18. Marcia Taylor, "Drought Renews Interest in Weather Policies," *Progressive Farmer*, Oct. 10, 2012, http://www.dtnprogressivefarmer.com/dtnag/common/link.do;jsessionid=9645814BE9A1EB8FDFA1CAF1EB2BE1BE.agfreejvm1?symbolicName=/free/news/template1&product=/ag/news/bestofdtnpf&vendorReference=0a68d0de-1353-41d7-a8d1-7272751ade39_1349786028506&paneContentId=88&paneParentId=0.
19. Christine Stebbins and Peter Bohan, "Top Farm Lender Worried by Drought, Politics," *Reuters*, Aug. 14, 2012, http://www.reuters.com/article/2012/08/14/us-usa-drought-lending-idUSBRE87D0VE20120814.

CHAPTER 5: UNITIES OF TIME AND SPACE

1. Andrew Leonard, "How Netflix Is Turning Viewers into Puppets," *Salon*, Feb. 1, 2013, http://www.salon.com/2013/02/01/how_netflix_is_turning_viewers_into_puppets.
2. Jehoshua Eliashberg et al., "Demand-Driven Scheduling of Movies in a Multiplex," *International Journal of Research in Marketing* 26, no. 2 (June 2009): 75–88, doi:10.1016/j.ijresmar.2008.09.004.
3. Barbara McManus, "Outline of Aristotle's Theory of Tragedy," Nov. 1999, CNR, http://www2.cnr.edu/home/bmcmanus/poetics.html.
4. Amy Kaufman, "Movie Attendance Falls to 16-Year Low," *Los Angeles Times*, Jan. 3, 2012, http://articles.latimes.com/2012/jan/03/entertainment/la-et-box-office-20120103.
5. Brent Harrison and David L. Roberts, "Using Sequential Observations to Model and Predict Player Behavior," in *Proceedings of the 6th International Conference on Foundations of Digital Games*, FDG 2011 (New York: ACM, 2011), 91–98, doi:10.1145/2159365.2159378.
6. Alexandra Alter, "Your E-Book Is Reading You," *Wall Street Journal*, July 19, 2012, http://online.wsj.com/article/SB10001424052702304870304577490950051438304.html.

7. J. E. Cutting, J. E. DeLong, and C. E. Nothelfer, "Attention and the Evolution of Hollywood Film," *Psychological Science* 21, no. 3 (Feb. 5, 2010): 432–39, doi:10.1177/0956797610361679.

CHAPTER 6 THE SPIRIT OF THE NEW

1. David Court et al., "The Consumer Decision Journey," *McKinsey Quarterly*, no. 3 (2009): 1–11.
2. Bill Lee, "Marketing Is Dead," *Harvard Business Review*, Aug. 9, 2012, http://blogs.hbr.org/cs/2012/08/marketing_is_dead.html.
3. Katherine Watts, "73% of CEOs Think Marketers Lack Business Credibility: They Can't Prove They Generate Business Growth," June 15, 2011, fournaisegroup.com, http://www.fournaisegroup.com/Marketers-Lack-Credibility.asp?_fwaHound=15127937_12185_15127937_0_0_0_0.
4. Joseph Turow, *The Daily You: How the New Advertising Industry Is Defining Your Identity and Your Worth* (New Haven: Yale University Press, 2011), 69.
5. Adrian Chen, "Jay-Z's New Album Is Basically a Massive Data-Mining Operation," *Gawker*, July 3, 2013, http://gawker.com/jay-zs-new-album-is-basically-a-massive-data-mining-op-661499440.
6. Karl Taro Greenfeld, "Loveman Plays 'Purely Empirical' Game as Harrah's CEO," Bloomberg, Aug. 6, 2010, http://www.bloomberg.com/news/2010-08-06/loveman-plays-new-purely-empirical-game-as-harrah-s-ceo.html.
7. "History of Loyalty Programs," FrequentFlier.com, accessed Jan. 21, 2013, http://www.frequentflier.com/programs/history-of-loyalty-programs.
8. "History of AMR and American Airlines," *American Airlines*, accessed Jan. 21, 2013, http://www.aa.com/i18n/amrcorp/corporateInformation/facts/history.jsp.
9. David Becker, "Gambling on Customers," *McKinsey Quarterly*, May 2003, http://www.forbes.com/2003/07/14/0714mckinsey.html.
10. Robert L. Shook, *Jackpot!: Harrah's Winning Secrets for Customer Loyalty* (Hoboken, NJ: Wiley, 2003), 235.
11. Meridith Levinson, "Harrah's Knows What You Did Last Night," *CIO*, June 6, 2001, http://www.cio.com.au/article/44514/harrah_knows_what_you_did_last_night.
12. Christina Binkley, "Taking Retailers' Cues, Harrah's Taps into Science of Gambling," *Wall Street Journal*, Nov. 22, 2004.
13. "Customer Analytics to Enable High Performance," *Accenture*, July 21, 2010, http://www.accenture.com/us-en/Pages/insight-customer-analytics-performance-summary.aspx.
14. "Wal-Mart Used Microchip to Track Customers," *WND*, accessed Jan. 21, 2013, http://www.wnd.com/2003/11/21809.
15. Thomas Davenport, *Realizing the Potential of Retail Analytics* (Babson College), SAS, accessed Jan. 21, 2013, http://www.sas.com/events/cm/622624/index.html.
16. V. Kumar et al., "Managing Retailer Profitability—One Customer at a Time!" *Journal of Retailing* 82, no. 4 (Dec. 2006): 277.

17. Aaron Smith, "In-store Mobile Commerce During the 2012 Holiday Shopping Season," *Pew Internet*, Jan. 31, 2013, http://www.pewinternet.org/Reports/2013/in-store-mobile-commerce.aspx.
18. Danica Kwon, "Survey Says: Most Americans Participate in Customer Loyalty Programs," *Polaris Point of View*, Aug. 6, 2012, http://www.polarismr.com/POV/bid/89305/Survey-Says-Most-Americans-Participate-in-Customer-Loyalty-Programs.
19. Chad Catacchio, "Walmart Now Doing Group Buying on Facebook," *Next Web*, Oct. 27, 2010, http://thenextweb.com/us/2010/10/27/walmart-now-doing-group-buying-on-facebook.
20. Candice Choi, "Companies Rate Consumers with New Scores," *Daily Herald*, Jan. 2, 2012.
21. "Walmart Corporate: Frequently Asked Questions," Walmart.com, accessed Jan. 22, 2013, http://stock.walmart.com/investor-faqs.
22. Michael Barbaro, "It's Not Only About Price at Wal-Mart," *New York Times*, Mar. 2, 2007, http://www.nytimes.com/2007/03/02/business/02walmart.html.
23. Acxiom, *Acxiom Product Briefing* (Palace Hotel, Acxiom Product Briefing, 2013), Shareholder.com.
24. Ibid.
25. Natasha Singer, "A Data Broker Offers a Peek Behind the Curtain," *New York Times,* accessed Oct. 17, 2013, http://www.nytimes.com/2013/09/01/business/a-data-broker-offers-a-peek-behind-the-curtain.html?ref=todayspaper&pagewanted=all&_r=0.
26. Eytan Bakshy et al., "The Role of Social Networks in Information Diffusion," Cornell University Library, Feb. 28, 2012, http://arxiv.org/abs/1201.4145.
27. Baldeesh Gakhal and Carl Senior, "Examining the Influence of Fame in the Presence of Beauty: An Electrodermal 'Neuromarketing' Study,'" *Journal of Consumer Behaviour: An International Research Review,* vol. 7.2008, 4/5, 331–41.
28. Jim Edwards, "How Facebook Will Reach $12 Billion in Revenue, Broken Down by Product," *Business Insider*, accessed Jan. 22, 2013, http://www.businessinsider.com/facebooks-annual-revenues-by-year-2012-12.
29. David Talbot, "Facebook's Latest Data Science Insight," *MIT Technology Review*, May 10, 2013, http://www.technologyreview.com/view/514796/facebooks-latest-data-science-insight.
30. Josh Constine, "Hands On with Facebook Nearby, a New Local Biz Discovery Feature That Challenges Yelp and Foursquare," *TechCrunch*, Dec. 17, 2012, http://techcrunch.com/2012/12/17/facebook-nearby.

CHAPTER 7: RELEARNING HOW TO LEARN

1. Maryjane Wraga et al., "Neural Basis of Stereotype-Induced Shifts in Women's Mental Rotation Performance," *Social Cognitive and Affective Neuroscience* 2, no. 1 (Mar. 1, 2007): 12–19, doi:10.1093/scan/nsl041.
2. "Digest of Education Statistics, 2011," Institute of Education Sciences, 2011, http://nces.ed.gov/programs/digest/d11/tables/dt11_008.asp.

3. Gillian Tett, "When Tablet Turns Teacher," *Financial Times*, Oct. 5, 2012, http://www.ft.com/cms/s/2/6a071e00-0db6-11e2-97a1-00144feabdc0.html#axzz2AWfQejIx.
4. Nicholas Negroponte, "EmTech Preview: Another Way to Think About Learning," *MIT Technology Review*, Sept. 13, 2012, http://www.technologyreview.com/view/429206/emtech-preview-another-way-to-think-about-learning.
5. John Gatto, "Against School," Sept. 2003, *Harper's Magazine*, http://harpers.org/archive/2003/09/against-school/.
6. "*Sugata Mitra: Kids Can Teach Themselves*" (LIFT, 2007), TED, http://www.ted.com/talks/sugata_mitra_shows_how_kids_teach_themselves.html.
7. "India—Hole in the Wall," Pbs.org, Oct. 2002, http://www.pbs.org/frontlineworld/stories/india/thestory.html.
8. Lee Felsenstein, "The Fonly Institute: Problems with the $100 Laptop," Nov. 10, 2005, *Fonly Institute*, http://www.fonly.typepad.com/fonlyblog/2005/11/problems_with_t.html.
9. Berk Ozler, "One Laptop Per Child Is Not Improving Reading or Math, but Are We Learning Enough from These Evaluations?" *Development Impact*, June 13, 2012, http://blogs.worldbank.org/impactevaluations/one-laptop-per-child-is-not-improving-reading-or-math-but-are-we-learning-enough-from-these-evaluations.
10. Sugata Mitra, R. Dangwai, and L. Thadani, "Effects of Remoteness on the Quality of Education: A Case Study from North Indian Schools," *Journal: Australasian Journal of Educational Technology* 24, no. 2 (2008): 168–80.
11. Patrick Tucker, "Futurist Magazine Update," vol. 13, no. 10, Oct. 2012, https://www.wfs.org/futurist-update/futurist-update-2012-issues/october-2012-vol-13-no-10.
12. Sue Holmes, "Sandia Shows Monitoring Brain Activity During Study Can Help Predict Test Performance," *Sandia Labs News Releases*, Sept. 10, 2012, https://share.sandia.gov/news/resources/news_releases/brain_study.
13. Ian Bogost, "Online Classes Won't Fix America's Unequal Education System," *Atlantic Cities*, Jan. 17, 2013, http://www.theatlanticcities.com/politics/2013/01/online-classes-wont-fix-americas-unequal-education-system/4424.

CHAPTER 8: WHEN YOUR PHONE SAYS YOU'RE IN LOVE

1. Paramahansa Yogananda, *Autobiography of a Yogi*, 11th ed. (Los Angeles: Self-Realization Fellowship, 1971).
2. Shardha Chettri, "Arranged Marriage in Hinduism—Its Point of Origin," *Family Journal* (Jan. 1, 2007).
3. Matthew Bramlett and William Mosher, *Cohabitation, Marriage, Divorce, and Remarriage in the United States*, Vital and Health Statistics, Series 23, No. 22 (Centers for Disease Control and Prevention, 2002).
4. "Calculating Match Percentages," okcupid.com, accessed Jan. 22, 2013, http://www.okcupid.com/help/match-percentages.

5. Jennifer Valentino-DeVries and Jeremy Singer-Vine, "They Know What You're Shopping For," *Wall Street Journal*, Dec. 7, 2012, http://online.wsj.com/article/SB10001424127887324784404578143144132736214.html.
6. Christian Rudder, "Why You Should Never Pay for Online Dating," OkTrends, Apr. 7, 2010, http://www.columbia.edu/~jhb2147/why-you-should-never-pay-for-online-dating.html.
7. Ibid.
8. Chris Coyne, "Rape Fantasies and Hygiene by State," OkTrends, Nov. 25, 2009, http://blog.okcupid.com/index.php/rape-fantasies-and-hygiene-by-state.
9. Eli J. Finkel et al., "Online Dating: A Critical Analysis from the Perspective of Psychological Science," *Psychological Science in the Public Interest* 13, no. 1 (Jan. 1, 2012): 3–66, doi:10.1177/1529100612436522.
10. Fritz Heider, "Attitudes and Cognitive Organization," *Journal of Psychology* 21, no. 1 (Jan. 1946): 107–12, doi:10.1080/00223980.1946.9917275.
11. R. Guha et al., "Propagation of Trust and Distrust" (presented at the WWW 2004, New York, 2004).
12. Jure Leskovec, Daniel Huttenlocher, and Jon Kleinberg, "Predicting Positive and Negative Links in Online Social Networks," 2010, 641–60, http://wwwconference.org/proceedings/www2010/www/p641.pdf.
13. Finkel et al., "Online Dating."
14. Yukari Iwatani, "Love: Japanese Style," *Wired*, June 11, 1998, http://www.wired.com/culture/lifestyle/news/1998/06/12899.
15. "Bleep at First Sight," *Reuters*, May 15, 1998, http://www.wired.com/culture/lifestyle/news/1998/05/12342.
16. Charlotte Kemp, "TAMAGOTCHIS THAT FIND A DATE: Is That a Lovegety in Your Pocket or Are You Just Pleased to See Me?" *Daily Mirror*, July 10, 1998, http://www.thefreelibrary.com/Is%20that%20a%20Lovegety%20in%20your%20pocket%20or%20are%20you%20just%20pleased%20to%20see%20me?; . . . -a060668653.
17. Nathan Eagle, *Machine Perception and Learning of Complex Social Systems* (Cambridge: Massachusetts Institute of Technology, 2005), http://realitycommons.media.mit.edu/pdfs/thesis.pdf.
18. Ibid.
19. Ibid.
20. "Match.com Releases Its Second Annual Comprehensive Study on Singles," Match.com, Feb. 2, 2012, http://match.mediaroom.com/index.php?s=43&item=116.
21. William Von Hippel and Robert Trivers, "The Evolution and Psychology of Self-deception," *Behavioral and Brain Sciences* 34 (2011): 1–56.
22. Alex Pentland, *Honest Signals: How They Shape Our World*, new ed. (Cambridge, MA: MIT Press, 2010).
23. Ibid., 28.
24. Ibid., 125.
25. Ibid., 132.
26. "Match.com Releases Its Second Annual Comprehensive Study on Singles," Match.com, Feb. 2, 2012.

27. "MIT Media Laboratory: Press Page—Jerk-O-Meter," accessed Jan. 22, 2013, http://www.media.mit.edu/press/jerk-o-meter.
28. Cindy Caldwell, "Preventing Catastrophic Incidents by Predicting Where They Are Most Likely to Occur and Why," Sept. 2012, http://semanticommunity.info/@api/deki/files/19429/Day2_1125_Caldwell_Larmey.pdf.
29. Josh Gold, "Workforce Instrumentation: The Art and Science of the Rational Enterprise" (presented at the O'Reilly Strata 2012, Santa Clara, CA, Feb. 28–Mar. 1), http://cdn.oreillystatic.com/en/assets/1/event/75/Business%20Intelligence_%20What%20have%20we%20been%20missing_%20Paper.pdf.
30. Reuben Hill, *Families Under Stress: Adjustment to the Crises of War Separation and Reunion*, rept. (Westport, CT: Greenwood Press, 1971).
31. Pentland, *Honest Signals*, 42.
32. Yogananda, *Autobiography of a Yogi*, 163.

CHAPTER 9: CRIME PREDICTION: THE WHERE AND THE WHEN

1. George L. Kelling and James Q. Wilson, "Broken Windows," *Atlantic*, Mar. 1982, http://www.theatlantic.com/magazine/archive/1982/03/broken-windows/304465.
2. Andreas Olligschlaeger, "Artificial Neural Networks and Crime Mapping," in *From Crime Mapping and Crime Prevention* (National Institute of Justice Drug Market Analysis Program United States, 1997), 313–47, https://www.ncjrs.gov/App/Publications/abstract.aspx?ID=170288.
3. Ibid., 342.
4. Jayne Marshall, *Zero Tolerance Policing* (Government of South Australia, Office of the Attorney General, Mar. 1999), http://www.ocsar.sa.gov.au/docs/information_bulletins/IB9.pdf.
5. Colleen McCue, *Data Mining and Predictive Analysis: Intelligence Gathering and Crime Analysis*, 1st ed. (Oxford: Butterworth-Heinemann, 2007).
6. Charlie Beck, "Predictive Policing: What Can We Learn from Wal-Mart and Amazon About Fighting Crime in a Recession?" *Police Chief Magazine*, Nov. 2009, http://www.policechiefmagazine.org/magazine/index.cfm?fuseaction=display_arch&article_id=1942&issue_id=112009.
7. Hanna Rosin, "American Murder Mystery," *Atlantic*, July–Aug. 2008, http://www.theatlantic.com/magazine/archive/2008/07/american-murder-mystery/306872.
8. Ingrid Ellen, Michael Lens, and Katherine O'Regan, *American Murder Mystery Revisited: Do Housing Voucher Households Cause Crime?* SSRN Scholarly Paper (Rochester, NY: Social Science Research Network, Mar. 6, 2012), http://papers.ssrn.com/abstract=2016444.
9. Jim Utsler, "The Crime Fighters," *IBM Systems Magazine*, Feb. 2011, http://www.ibmsystemsmag.com/power/trends/ibmresearch/ibm_research_spss.
10. Matthew Freedman and Emily Owens, "Your Friends and Neighbors: Localized Economic Development, Inequality, and Criminal Activity," 2012, http://works.bepress.com/matthew_freedman/17.
11. Utsler, "The Crime Fighters."

12. "NYPD and Microsoft Build Hi-tech Crime Fighting 'Dashboard,'" *Telegraph*, Feb. 20, 2013, http://www.telegraph.co.uk/technology/news/9884479/NYPD-and-Microsoft-build-hi-tech-crime-fighting-dashboard.html.
13. *CCTV Based Remote Biometric and Behavioral Suspect Detection: Technologies and Global Markets—2011–2016, Homeland Security Market Research* (Homeland Security Research, Q1 2011), http://www.homelandsecurityresearch.com/2013/08/china-homeland-security-public-safety-market-2013-edition; for more details https://www.wfs.org/futurist/2013-issues-futurist/march-april-2013-vol-47-no-2/chinas-closed-circuits.
14. John L. Rep Mica, *H.R.658—FAA Modernization and Reform Act of 2012*, 2011, http://thomas.loc.gov/cgi-bin/bdquery/z?d112:h.r.658.
15. "Subcommittee Hearing: Using Unmanned Aerial Systems Within the Homeland: Security Game Changer?" accessed Jan. 22, 2013, http://homeland.house.gov/hearing/subcommittee-hearing-using-unmanned-aerial-systems-within-homeland-security-game-changer.
16. William Bratton, *Zero Tolerance: Policing a Free Society*, IEA Health and Welfare Unit: Choice in Welfare, No. 35 (IEA Health and Welfare Unit London, Apr. 1997), http://www.civitas.org.uk/pdf/cw35.pdf.
17. Marta Molina, "Yo Soy 132 Rejects Election Results, Continues Organizing," *Indy Blog*, July 13, 2012, http://www.indypendent.org/2012/07/13/yo-soy-132-rejects-election-results-continues-organizing.
18. Feng Chen et al., *Spatial Surrogates to Forecast Social Mobilization and Civil Unrests1* (Virginia Tech, 2012), http://people.cs.vt.edu/~naren/papers/CCC-VT-Updated-Version.pdf.
19. David Joachim, "What Is Intelligence Chatter, Anyway?" *Slate Explainer*, Sept. 12, 2003, http://www.slate.com/articles/news_and_politics/explainer/2003/09/what_is_intelligence_chatter_anyway.html.
20. *Making the World a Witness: Report on the Pilot Phase, Satellite Sentinel Project*, Dec. 2010, http://www.satsentinel.org/report/making-world-witness-report-pilot-phase-report.
21. Andrew Zammit-Mangion et al., "Point Process Modelling of the Afghan War Diary," *Proceedings of the National Academy of Sciences* (July 16, 2012), doi:10.1073/pnas.1203177109.
22. Kalev Leetaru, "Culturomics 2.0: Forecasting Large-scale Human Behavior Using Global Media Tone in Time and Space," *First Monday* 16, no. 9 (Sept. 2011), http://firstmonday.org/htbin/cgiwrap/bin/ojs/index.php/fm/article/view/3663/3040.
23. Marco Lagi, Karla Z. Bertrand, and Yaneer Bar-Yam, "The Food Crises and Political Instability in North Africa and the Middle East," *New England Complex Systems Institute* (July 19, 2011).

CHAPTER 10: CRIME: PREDICTING THE WHO

1. Adam Higginbotham, "Deception Is Futile When Big Brother's Lie Detector Turns Its Eyes on You," *Wired*, Feb. 2013, http://www.wired.com/threatlevel/2013/01/ff-lie-detector.
2. Ralph Chatham, "Ralph Chatham's Informal Summary of the Insights and Findings from DARPA's Rapid Checkpoint Screening Program," Jan. 2007,

https://www.google.com/url?sa=t&rct=j&q=&esrc=s&source=web&cd=1&ved=0CDIQFjAA&url=http%3A%2F%2Fwww.cl.cam.ac.uk%2F~rja14%2Fshb08%2Fchatham1.doc&ei=1AT_UPamCIyq0AHoioHwCQ&usg=AFQjCNFzJAlk-NVHrcb6DcE9SRnREq6NWQ&sig2=21WFlClt-ekMk7qmlCAFsw&bvm=bv.41248874,d.dmQ.

3. I. Pavlidis, J. Levine, and P. Baukol, "Thermal Image Analysis for Anxiety Detection," in *2001 International Conference on Image Processing, 2001, Proceedings*, vol. 2, 2001, 315–18, doi:10.1109/ICIP.2001.958491.
4. *Hearing on Behavioral Science and Security: Evaluating TSA's SPOT Program*, 2011, http://science.house.gov/sites/republicans.science.house.gov/files/documents/hearings/2011%2003%2030%20Ekman%20Testimony.pdf.
5. Michael Kimlick, *Privacy Impact Assessment for the Screening of Passengers by Observation Techniques (SPOT) Program* (Washington, DC: U.S. Department of Homeland Security, Aug. 5, 2008), http://www.dhs.gov/xlibrary/assets/privacy/privacy_pia_tsa_spot.pdf.
6. Eric Lipton, "Faces, Too, Are Searched at U.S. Airports," *New York Times*, Aug. 17, 2006, http://www.nytimes.com/2006/08/17/washington/17screeners.html.
7. Anthony Kimery, "TSA's SPOT Program Not Scientifically Grounded GAO Told Congress: TSA, Experts Disagree," *Homeland Security Today*, Apr. 7, 2011, http://www.hstoday.us/briefings/daily-news-briefings/single-article/tsa-s-spot-program-not-scientifically-grounded-gao-told-congress-tsa-experts-disagree/66b9300d981c1b1a39ac475411d38739.html.
8. "Majority Views NSA Phone Tracking as Acceptable Anti-Terror Tactic," *Pew Research Center for the People and the Press*, June 10, 2013, http://www.people-press.org/2013/06/10/majority-views-nsa-phone-tracking-as-acceptable-anti-terror-tactic.
9. Scott Huddleston, "Hasan Sought Gun with 'High Magazine Capacity,'" *My San Antonio*, Oct. 21, 2010, http://blog.mysanantonio.com/military/2010/10/hasan-sought-gun-with-high-magazine-capacity.
10. James C. McKinley Jr. and James Dao, "Fort Hood Gunman Gave Signals Before His Rampage," *New York Times*, Nov. 9, 2009, http://www.nytimes.com/2009/11/09/us/09reconstruct.html.
11. Christine Baker, "A Change of Detection: To Find the Terrorist Within the Identification of the U.S. Army's Insider Threat" (U.S. Army Command and General Staff College), accessed Jan. 22, 2013, http://www.hsdl.org/?view&did=723130.
12. "Executive Order 13587—Structural Reforms to Improve the Security of Classified Networks and the Responsible Sharing and Safeguarding of Classified Information," Oct. 7, 2011, http://www.whitehouse.gov/the-press-office/2011/10/07/executive-order-structural-reforms-improve-security-classified-networks.
13. "DARPA—Anomaly Detection at Multiple Scales (ADAMS)," Oct. 19, 2010, *Scribd*, http://www.scribd.com/doc/40392649/DARPA-Anomaly-Detection-at-Multiple-Scales-ADAMS.
14. Robert H. Anderson and Richard Brackney, *Understanding the Insider Threat* (RAND Corporation, 2004), http://www.rand.org/pubs/conf_proceedings/CF196.html.

15. Claire Cain Miller, "Tech Companies Concede to Surveillance Program," *New York Times*, June 7, 2013, http://www.nytimes.com/2013/06/08/technology/tech-companies-bristling-concede-to-government-surveillance-efforts.html.
16. Baker, "A Change of Detection."
17. Ibid.
18. O. Brdiczka et al., "Proactive Insider Threat Detection Through Graph Learning and Psychological Context," in *2012 IEEE Symposium on Security and Privacy Workshops (SPW)*, 2012, 142–49, doi:10.1109/SPW.2012.29.
19. Nate Berg, "Want to Shame a Terrible Parker? There's an App for That," *Atlantic Cities*, May 21, 2012, http://www.theatlanticcities.com/technology/2012/05/want-report-terrible-parker-theres-app/2055.
20. "Mining Social Networks: Untangling the Social Web," *Economist*, Sept. 2, 2010, http://www.economist.com/node/16910031.
21. Julia Angwin, "U.S. Terrorism Agency to Tap a Vast Database of Citizens," *Wall Street Journal*, Dec. 13, 2012, http://online.wsj.com/article/SB10001424127887324478304578171623040640006.html?user=welcome&mg=id-wsj.
22. David Thissen and Howard Wainer, "Toward the Measurement and Prediction of Victim Proneness," *Journal of Research in Crime and Delinquency* 20, no. 2 (July 1, 1983): 243–61, doi:10.1177/002242788302000206.

CHAPTER 11: THE WORLD THAT ANTICIPATES YOUR EVERY MOVE

1. Mike Orcutt, "The Pressure's on for Intel," *MIT Technology Review*, Nov. 9, 2012, http://www.technologyreview.com/news/507011/the-pressures-on-for-intel.
2. Jeff Hawkins and Sandra Blakeslee, *On Intelligence*, 1st ed. (New York: Times Books, 2004), 6.
3. Jason P. Gallivan et al., "Decoding Action Intentions from Preparatory Brain Activity in Human Parieto-Frontal Networks," *The Journal of Neuroscience* 31, no. 26 (June 29, 2011): 9599–610, doi:10.1523/JNEUROSCI.0080-11.2011.
4. Moshe Bar, "The Proactive Brain: Memory for Predictions," *Philosophical Transactions of the Royal Society B: Biological Sciences* 364, no. 1521 (May 12, 2009): 1235–43, doi:10.1098/rstb.2008.0310.
5. Ibid.
6. Scott A. Huettel, Peter B. Mack, and Gregory McCarthy, "Perceiving Patterns in Random Series: Dynamic Processing of Sequence in Prefrontal Cortex," *Nature Neuroscience* 5, no. 5 (May 2002): 485–90, doi:10.1038/nn841.
7. David Brin, *The Transparent Society: Will Technology Force Us to Choose Between Privacy and Freedom?* 1st trade paper ed. (New York: Basic Books, 1999).

INDEX